Seabed Minerals Series
Volume 4

Delineation of Mine-Sites and Potential in Different Sea Areas

Seabed Minerals Series

Volume 1. **Assessment of Manganese Nodule Resources**

Volume 2. **Analysis of Exploration and Mining Technology for Manganese Nodules**

Volume 3. **Analysis of Processing Technology for Manganese Nodules**

Volume 4. **Delineation of Mine Sites and Potential in Different Sea Areas**

Volume 5. **Selection of Sites for Nodule Processing Plants**

Volume 6. **Financial Analysis of a Manganese Nodule Mining Venture: A Balance Sheet**

Volume 7. **Economics of Lateritic Nickel Production in Developing Countries**

Volume 8. **Regulatory Framework for Manganese Nodule Development**

Volume 9. **Review of Activities of International Consortia and other Public and Private Entities**

Seabed Minerals Series
Volume 4

Delineation of Mine-Sites and Potential in Different Sea Areas

United Nations
Ocean Economics and Technology Branch

Published in co-operation with the United Nations
by Graham & Trotman Limited

Published in 1987 by

Graham & Trotman Limited
Sterling House
66 Wilton Road
London SW1V 1DE

Graham & Trotman Inc.
101 Philip Drive
Assinippi Park
Norwell
MA 02061
USA

in co-operation with the United Nations

Softcover reprint of the hardcover 1st edition 1987

ISBN-13: 978-94-010-7952-5 e-ISBN-13: 978-94-009-3235-7
DOI: 10.1007/978-94-009-3235-7

Typeset in Great Britain by Bookworm Typesetting, Manchester

Contents

Acknowledgements

We are grateful to J.P. Lenoble and D. Pasho for their valuable contributions to this publication and also wish to express our appreciation to A.A. Archer, J.Z. Frazer, and V.E. McKelvey for their many helpful comments and suggestions. We also acknowledge with thanks the assistance of G. Zahariev for preparation of the material on statistical methods.

Introduction

Prospecting and exploration for manganese nodules has, as its ultimate objective, the discovery and delineation of an area of the ocean floor with reserves of sufficient quantity and quality to support a mining operation under existing economic, technical and political conditions. While prospecting concentrates primarily on the collection of geological information, an exploration programme includes other activities that relate to the development of technology, financial analysis of the prospect and environmental protection. Such work on a deposit in turn leads to the development of a mine-site.

The mine-site concept brings together information in a way that recognizes the interplay among a number of dynamic factors which must satisfy a set of technical and economic conditions. Defining a mine-site, therefore, is a process of accounting for those factors.

Throughout the years of meetings of the Third United Nations Conference on the Law of the Sea, many questions arose about ocean mine-sites. Two related topics in particular received attention: the total number of available mine-sites, and the amount of area necessary for a mining operation. Both of these topics have been subject to a great deal of speculation, and even with the best available information, there remains a degree of uncertainty that arises from both incomplete knowledge and natural variability in the seabed and the resource, and different technology and production objectives. For example, estimates of the size of the area necessary for an ocean mine-site vary even when made by the same company.

On 15 November 1974, Deepsea Ventures, Inc., representing one of the ocean mining consortia (see Appendix A for identification of entities involved in mine-site development), filed a "Notice of Discovery and Claim of Exclusive Mining Rights, and Request for Diplomatic Protection of Investment" with the United States Department of State. In this document, the location and extent of the first ocean claim intended for the commercial production of manganese nodules was defined as follows:

> From: (1) Latitude 15°44′ N, Longitude 120°20′ W. A line drawn West to: (2) Latitude 15°44′ N, Longitude 127°46′ W. And thence South to: (3) Latitude 14°16′ N, Longitude 127°46′ W. And thence East to: (4) Latitude 14°16′ N, Longitude 124°20′ W. And thence North to the point of origin.

> These lines include approximately 50,000 square kilometres for purpose of development and evaluation of the Deposit encompassed therein, which area will be reduced by Deepsea to 30,000 square kilometres upon expiration of a term of 15 years (absent force majeure) from the date of this notice or upon commencement whichever occurs first. The Deposit lies on the abyssal ocean floor in water depths ranging between 2,300 to 5,000 metres and is more than 1,000 kilometres from the nearest island, and more than 1,300 kilometres seaward of the outer edge of the nearest continental margin. It is beyond the limit of seabed jurisdiction presently claimed by any State.
>
> (Deepsea Ventures, 1974)

Ten years later, in 1984, Deepsea Ventures Inc filed a final amendment to its application for exclusive exploration rights under United States legislation. This amendment indicated that 156,000 rather than 50,000 square kilometres was needed for exploration.

This increase in area reflects changes in the technical, economic and political assumptions made by the mining industry over a period of about a decade.

During this period a considerable amount has been written on the subject of ocean mine-sites. This volume examines the mine-site concept as it has evolved and evaluates the reasons for the wide variation in estimates of both the number of available sites and in estimates of the size of the area necessary for commercial production. Chapter 1 reviews the nature, distribution and extent of manganese nodule resources and includes a summary of areas of the World Ocean with potential resources. Chapter 2 reviews the economic, technical and geological factors used to define manganese nodule reserves and describes the methods used to determine sizes of mine-sites. Chapter 3 reviews estimates of mine-site availability. Chapter 4 discusses the exploration area, its relationship to the mine-site size and describes the existing areas for which the United States has issued licences. Chapter 5 describes mine-site configuration, regulatory approaches toward mine-site size and shape and the implications of these approaches. Chapter 6 presents conclusions.

Chapter 1

MANGANESE NODULE RESOURCES: Nature, Distribution and Extent

Manganese nodule deposits have been described extensively in the literature and in the preceeding volumes of this series. Characteristics noted here are those which have implications for mine-site definition. Manganese nodules are concretionary bodies that accumulate *in situ* in a single layer on the surface of the seabottom. This physical configuration of the deposit means that a manganese nodule mine-site will extend great distances in two directions. Based on commonly accepted assumptions, an ocean mine-site could cover an area as small as 30,000 km^2 (the size of Belgium) or as large as 100,000 km^2 (the size of Jordan).

It is generally agreed that no single formation process is responsible for nodule genesis. Because several processes are at work, it is to be expected that there will be considerable variation in their chemical and physical properties.

Manganese nodules are widely distributed throughout the World Ocean (Fig. 1). The most favourable areas are those with low sedimentation rates, remote from land and in deep water. The most extensive nodule fields are found in the Pacific Ocean. The Atlantic Ocean has higher rates of terrigenous sedimentation and shallower average water depths; accordingly, the percentage of seafloor covered with nodules is small. The Indian Ocean has fewer nodules than the Pacific, but more than the Atlantic.

Maps of worldwide distribution of nodules have little meaning without information on abundance (weight of nodules per unit area of seafloor, usually expressed as kg/m^2). Nodule abundance is variable. Closely spaced sampling and photography have revealed that nodule abundance varies considerably over distances of a kilometre or less. Thus there are great difficulties in estimating nodule abundance, even in those areas where photographs and bottom samples have been taken together. These

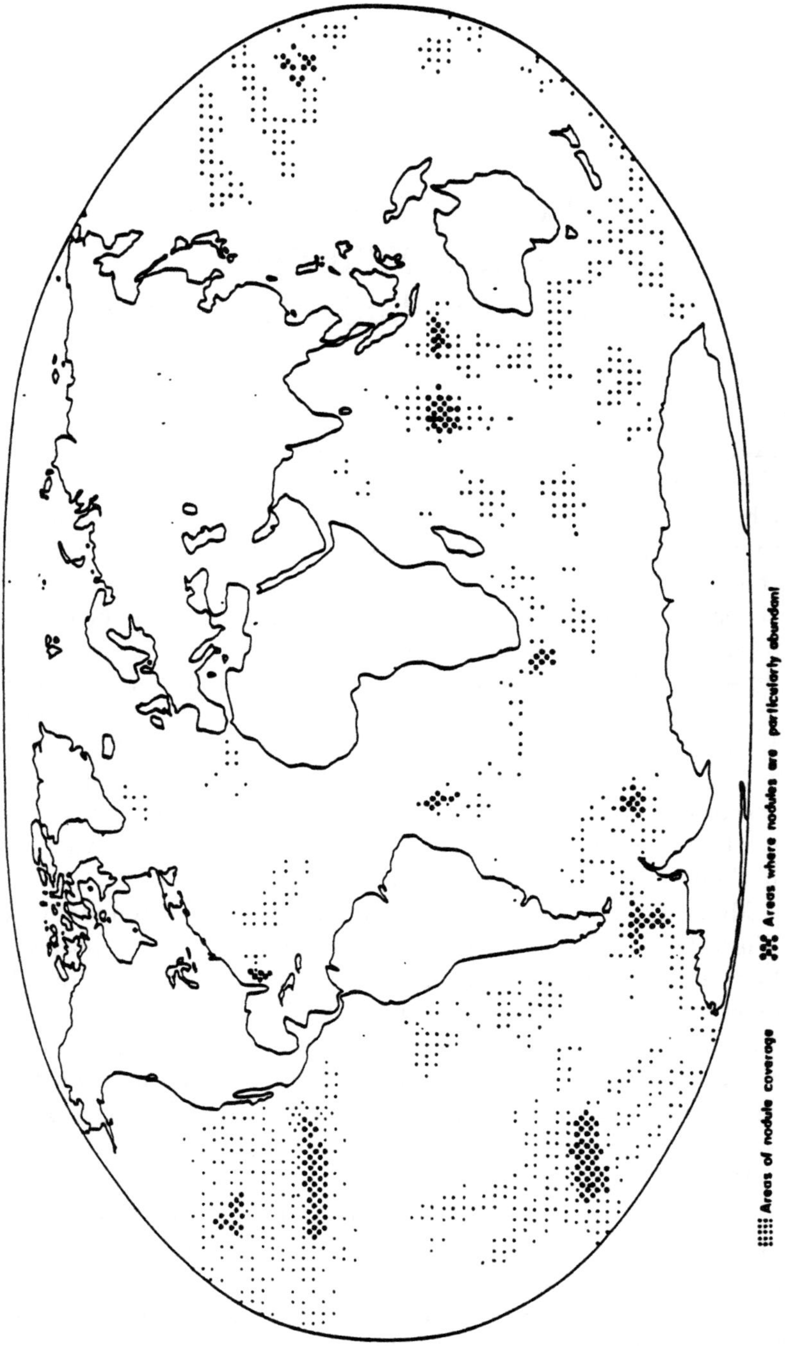

Fig. 1. Distribution of manganese nodules in the World Ocean (after Rawson and Ryan 1978, and other sources). (D.S. Cronan, *Underwater Minerals,* 1980, Fig. 32, p.91. Reprinted with permission from Academic Press, London).

variations, coupled with changes in chemical composition, contribute to the uncertainty involved in resource estimation and mine-site selection.

Another difficulty relates to the problems of exploring the resource itself. The nodule fields of greatest commercial interest lie at a depth of about 5000 metres, far from land. Surveying is complex and costly; measurements are often indirect. General studies concerning the geology, geochemistry, and mineralogy of manganese nodules have contributed much of the publicly available information used to define a commercial mine-site. From a commercial viewpoint, such studies have shortened exploration and contributed to predicting where the richest and most abundant resources lie and decreased the time and expense of exploration.

RESOURCE ASSESSMENT

There are two basic classifications used to describe the significance of mineral deposits: reserves and resources. Mineral reserves are those mineral occurrences that are exploitable under prevailing economic conditions. Mineral resources are those mineral occurrences that cannot be economically recovered under prevailing economic conditions, but that may be exploitable in the foreseeable future. Further subdivisions exist within both categories as indicated in the US Department of Interior's classification chart for mineral resources (Fig. 2). These subdivisions are predicated on two classes of information: geological and economic. It is important to note that resources which are uneconomic to exploit at any particular time can become economic at a later date when the price and market conditions, or new technologies, increase the profit margin to acceptable levels. At this time there are no identified manganese nodule reserves under the terms of this definition, only subeconomical paramarginal resources.

The process of resource assessment is theoretically a simple series of steps. For any given mineral, once it has been discovered, a series of measurements are taken by which the mineral deposit is characterized. Then a set of criteria is developed from an analysis of production requirements. The deposit is then classified as a resource or reserve. The result is that a mineral deposit is given significance by combining

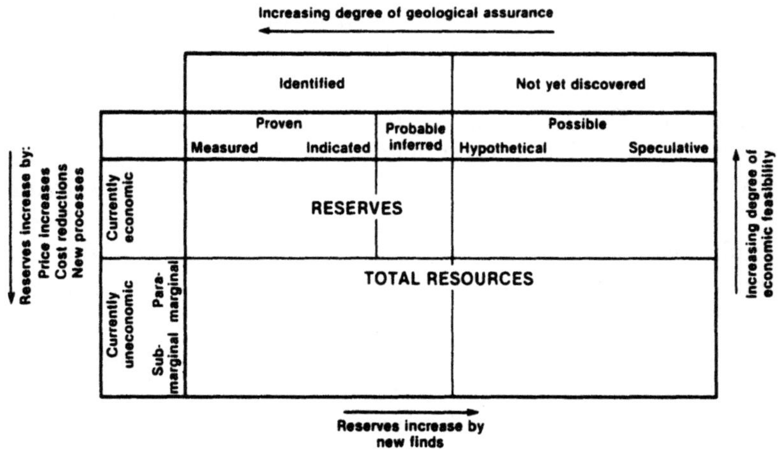

Fig. 2. US Department of Interior classification system for mineral resources.

geological knowledge and economic knowledge. For certain types of marine minerals, the process is in fact, very difficult. For example, in deep ocean mineral development, it is more than likely that the available geological information might be extremely limited. In the past, areas with potential resources have been identified on the basis of one or two samples which are taken to represent ore lying over thousands of square kilometres of ocean floor. On the economic side, one is faced with the development of new technology, the costs and efficiencies of which are uncertain. The uncertainties in making resource estimates therefore, cannot be emphasized too strongly in the case of deep ocean mineral development.

In spite of the problems mentioned above, the task of exploration and subsequent mine-site definition would be far more difficult if past experience with land-based exploration and mining could not be applied. In fact, there are many similarities between land and ocean mining development projects, particularly with regard to the general strategy of exploration. Some of these similarities have been pointed out by Kaufman (1974). He states that: (1) both call for sufficient mineralization in a host rock or sediment to constitute an ore body; (2) both start with large areas from which unpromising areas are excluded; (3) both require decisions, based on the odds for success in the remaining tracts, as to how much to invest; and (4) both end up with a tract, or tracts within the areas of interest, that hopefully contain an ore body of interest. He also points out that neither land nor ocean miners can finally prove the exact extent of an ore body until after mining is completed.

The criteria for assessing manganese nodule resources whether it be regionally, globally, or for relatively small areas, are the same. Geological parameters such as sediment type, sedimentation rate, current activity, element assay and the density of nodules on the sea floor give a rough idea about the probability of finding potentially economic nodules in an area. But geological indicators alone do not determine whether such an occurrence is classified as a resource or as a reserve. The criteria that have been developed for determining whether a nodule occurrence can be defined as representing a resource or a reserve are dependent upon the mining system, the area's mineral endowment and characteristics, supply and demand for the contained metals and the assumed profitability of the specific ocean mining operation.

The first criterion for identifying resources is *adequate grade.* Because revenues from a manganese nodule mining operation will be generated from the sale of more than one metal, a definition of adequate grade should reflect the polymetallic characteristic of the ore. It has been suggested that because nickel and copper sales will provide the major portion of the revenues for most ocean mining operations, grade can be expressed as the sum of the nickel plus copper percentages in the same nodule. In 1978 the United Nations held a Group of Experts meeting attended by both industry representatives and academicians (United Nations, 1979). It was their conclusion that, at that time, the minimum acceptable average grade required to qualify manganese nodules as potential reserves was about 2.25% combined nickel and copper. The lowest average grade required to qualify manganese nodules as resources was about 1.5% nickel plus copper.

The second criterion is *adequate abundance.* At the United Nations meeting, the lowest acceptable average abundance required to qualify manganese nodules as potential reserves was estimated at about 10 kg/m^2 (wet weight), and at about 5 kg/m^2 (wet) to qualify manganese nodules as resources.

PROSPECTED AREAS

It has been known for more than a decade that the northeastern equatorial Pacific is the area with the greatest commercial potential for manganese nodule mining. Other areas of the World Ocean have been investigated; however, no other area has been found to have the combined grade and abundance of nodules as that of the North Pacific.

Target areas are those areas that appear to contain metal-rich nodules of high enough quality to justify further exploration. Several researchers have attempted to pinpoint target areas using published data as the basis for their analyses. In addition to this desk-top approach, several nations, and at least one regional international organization, have conducted prospecting cruises in different areas of the Ocean. The following section will summarize generally the areas of the World Ocean that have been investigated in terms of meeting or not meeting the criteria necessary to justify further exploration. Appendix A contains definitions of terms used in the following discussions.

Early assessments based on public data took the form of compilations of chemical data from the literature and were presented as tables and maps. This work was supported by the United States National Science Foundation as part of the International Decade of Ocean Exploration (IDOE). Even with very little data, some of it of questionable quality, the North Pacific Ocean nodules were shown to have higher manganese, nickel and copper contents than nodules in other oceans. As sampling increased, the development of computerized data banks of nodule chemical data and computer mapping became a necessity. The Scripps Institution of Oceanography's World Ocean Sediment Data Bank (Frazer and Fisk, 1978) has served as the main compilation and source of chemical data for almost every researcher in the field of manganese nodule resource assessment. The latest and most comprehensive survey of this type was done by McKelvey *et al.* in 1982. The selection of target areas for first-generation mining was based on a cut-off grade of 1.8 percent combined nickel and copper. Other factors important to mine-site selection such as abundance and topography were not taken into consideration. However, the main thrust of the paper was to locate areas deserving further study. McKelvey's maps are reproduced here, for the purpose of information, as Figs 3(a) to 3(j).

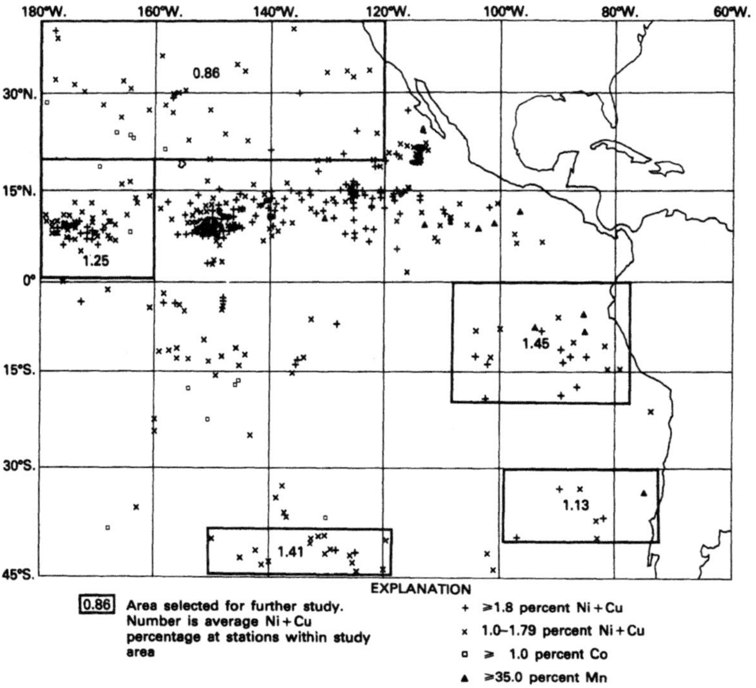

Fig. 3(a). Distribution in the Pacific Ocean of manganese nodules having high metal contents. Reprinted from McKelvey *et al.*, 1982 (Fig. 19).

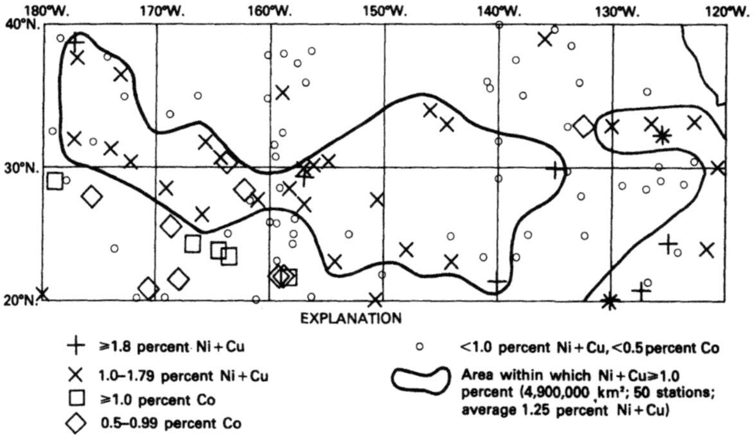

Fig. 3(b). Target areas for exploration for manganese nodules rich in combined nickel and copper in the central northeastern Pacific. Reprinted from McKelvey *et al.*, 1982 (Fig. 22).

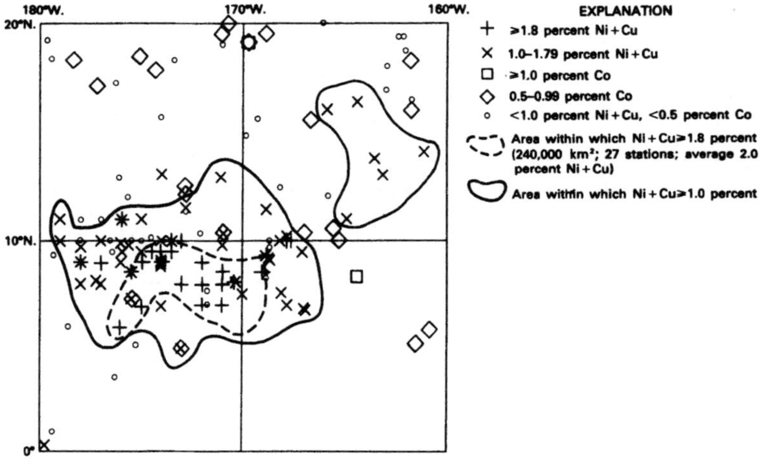

Fig. 3(c). Target areas for exploration for manganese nodules rich in combined nickel and copper in the central north equatorial Pacific. Reprinted from McKelvey *et al.,* 1982 (Fig. 23).

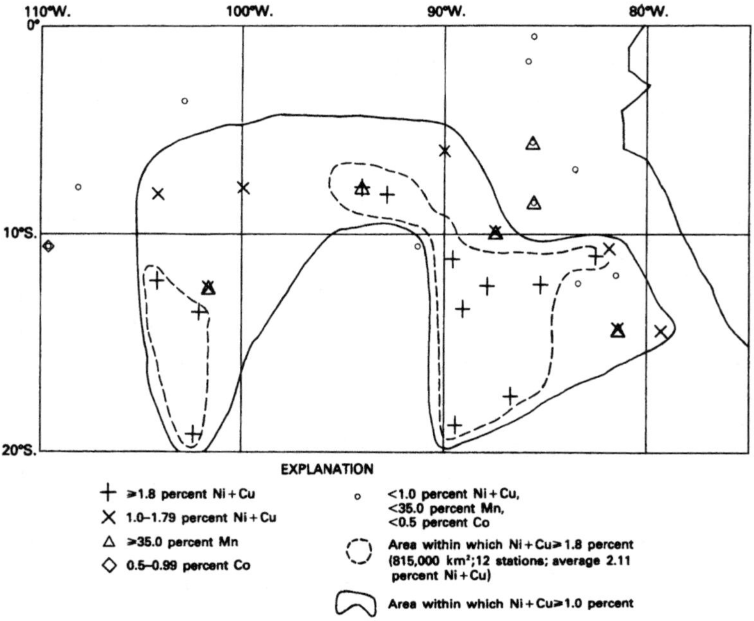

Fig. 3(d). Target areas for exploration for manganese nodules rich in combined nickel and copper in the southeastern equatorial Pacific. Reprinted from McKelvey *et al.,* 1982 (Fig. 24).

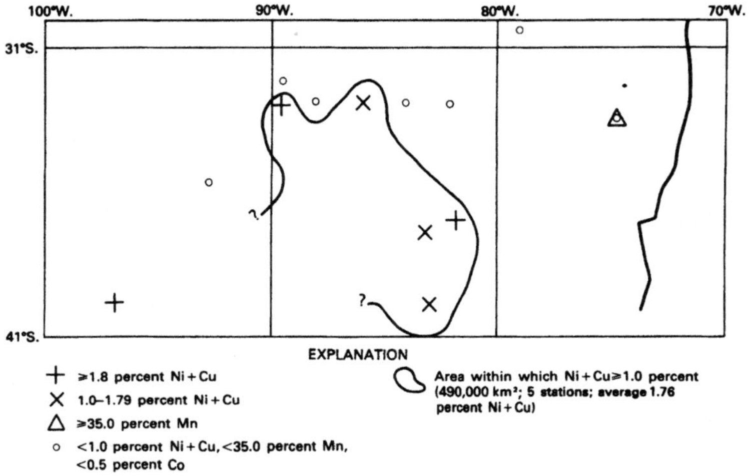

Fig. 3(e). Target areas for exploration for manganese nodules rich in combined nickel and copper in the central southeastern Pacific. Reprinted from McKelvey *et al.*, 1982 (Fig. 25).

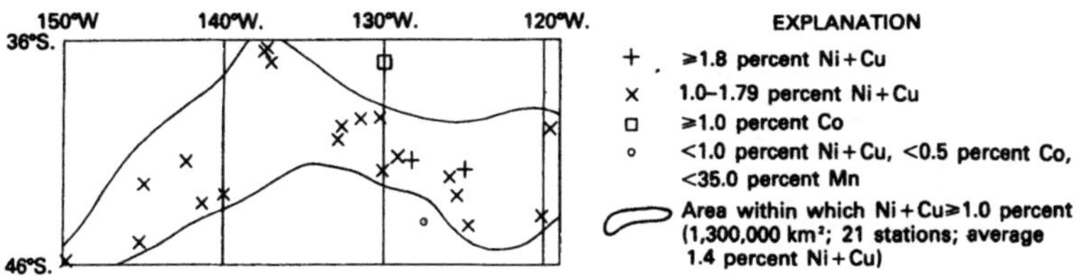

Fig. 3(f). Target areas for exploration for manganese nodules rich in combined nickel and copper in the central South Pacific. Reprinted from McKelvey *et al.*, 1982 (Fig. 26).

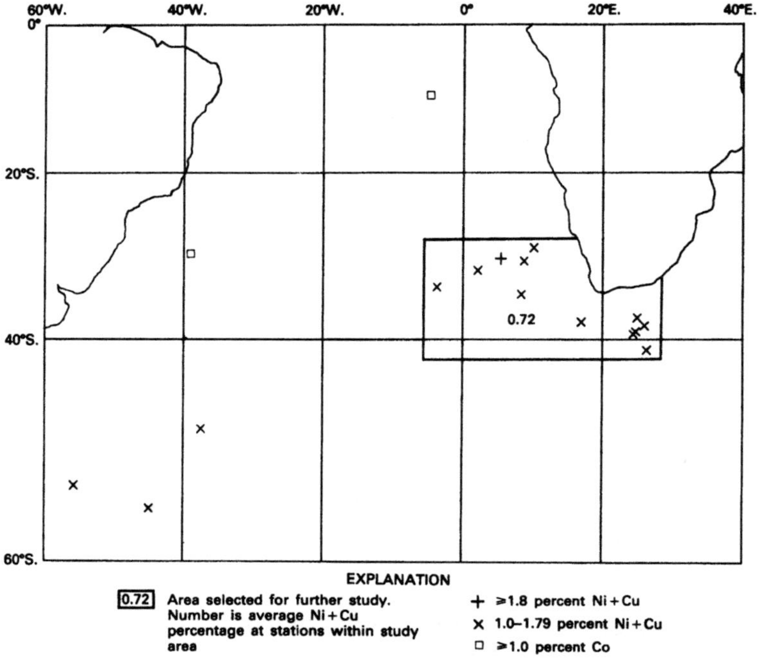

Fig. 3(g). Distribution in the South Atlantic of manganese nodules having high metal contents. Reprinted from McKelvey *et al.*, 1982 (Fig. 20).

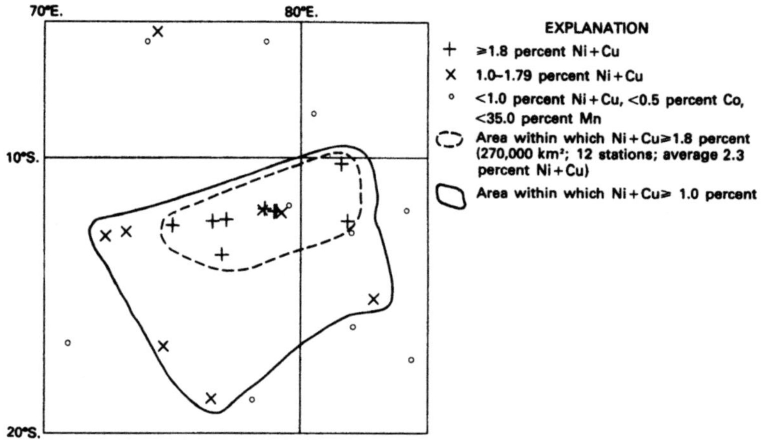

Fig. 3(i). Target areas for exploration for manganese nodules rich in combined nickel and copper in the central south equatorial Indian Ocean. Reprinted from McKelvey *et al.*, 1982 (Fig. 27).

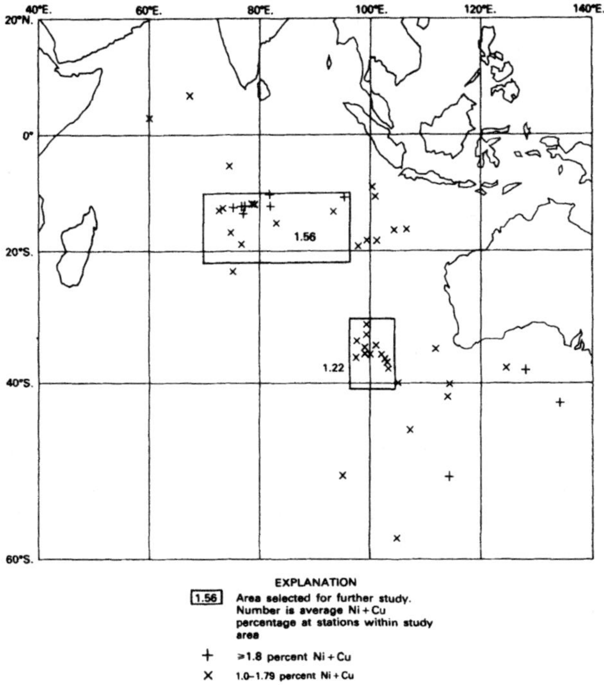

Fig. 3(h). Distribution in the Indian Ocean of manganese nodules having high metal contents. Reprinted from McKelvey, *et al.*, 1982 (Fig. 21).

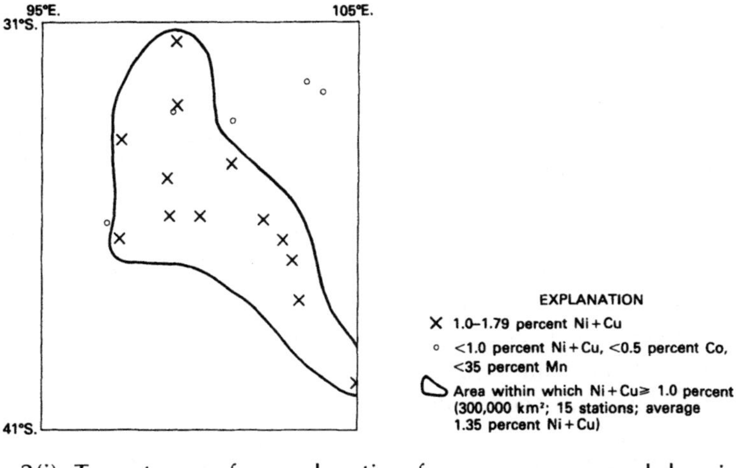

Fig. 3(j). Target areas for exploration for manganese nodules rich in combined nickel and copper in the southeastern Indian Ocean. Reprinted from McKelvey *et al.*, 1982 (Fig. 28).

Many of the areas mentioned in early reports have since been prospected by the mining industry or organizations engaged in prospecting for mineral resources. A brief summary of those areas follows.

PROSPECTED AREAS OUTSIDE THE CLARION–CLIPPERTON ZONE

Sites of interest to the mining industry were suggested by Horn *et al.* in 1973. The conclusions reached by the authors at that time were that the low metal contents of all deposits of the Atlantic, Indian and South Pacific Oceans rendered those oceans of little interest to the ocean mining community. Although it is true that the area of principal interest remains the North Pacific, the ocean mining industry has spent considerable time and money in exploration of several areas outside the North Pacific.

Atlantic Ocean

The Atlantic Ocean has held minimal interest for nodule miners due to the consistently low metal values found in the nodules there. Low assay values combined with other negative geological indicators have sustained the early conclusion that no deposits of current economic interest will be found.

Indian Ocean

Prospecting by the RV *Valdivia* supported by both government and commercial groups in the Federal Republic of Germany took place in the Indian Ocean in 1973 and 1974 (Glasby, 1984) (Fig. 4, Area A).

Five regions of the Indian Ocean were assessed by Frazer and Wilson (1980) using available data, some of which came from the *Valdivia* cruises, on nodule grade and abundance. The criteria used were an average of 2.4 percent combined cobalt, copper and nickel, with a cut-off grade of 1.8 percent and cut-off abundance of 5 kg/m^2. The paper concluded that only the Central Indian Ocean Basin offered possible sites for first generation mining and suggested further that it was extremely unlikely that mineable deposits would be found outside the areas studied.

Additional prospecting of the Indian Ocean took place in 1981

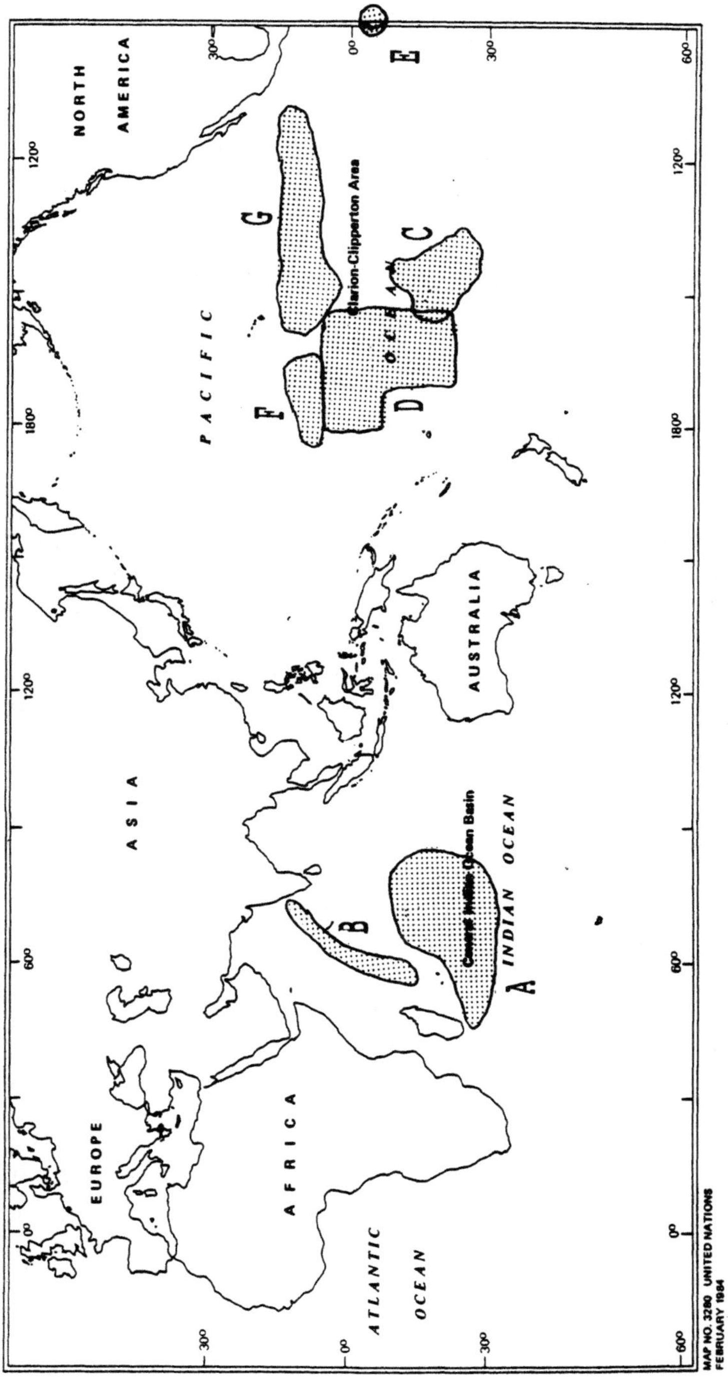

Fig. 4. Prospected Areas of the World Ocean.

by the RV *Gaveshani* under the auspices of the National Institute of Oceanography, Goa, India (Fig. 4, Area B). Information on subsequent research by the *Gaveshani* is scarce but the Indian government has requested registration of a mine-site area with the Preparatory Commission for the International Sea-Bed Authority and for the International Tribunal for the Law of the Sea, so it must be assumed that the results of the prospecting cruises were encouraging enough to warrant additional investment in exploration.

South Pacific

Between 1970 and 1981, the Centre National pour l'Exploitation des Océans (CNEXO) undertook 13 cruises in the South Pacific. These cruises were all conducted in the area of French Polynesia (Fig. 4, Area C). The CNEXO surveys showed that the overall abundance was low (50 per cent of the stations sampled had less than 1 kg/m^2), and in areas where abundance was high (9 kg/m^2) the nodules were low in nickel and copper content. The cut-off grade (Ni + Cu) was 1.7 percent. In those areas where the grade was above cut-off, the abundance averaged 2.5 kg/m^2 (ESCAP/CCOP/SOPAC, 1981). In 1974, CNEXO took a major decision to abandon offshore French Polynesia and concentrate their efforts in the North Pacific. They gave the following reasons for this decision: "Prospecting around Polynesia revealed, in effect, both a complex and very heterogeneous geographical distribution and an inverse correlation between grade and abundance". However, "without barring an eventual interest in the Polynesian area, and in any case, the French interest having been safeguaded, it appeared more reasonable to concentrate efforts in the north" (l'Académie des Sciences, 1984 — unofficial translation). As a result of this policy, the considerable amount of data collected by CNEXO has remained confidential.

The Committee for Co-ordination of Joint Prospecting for Mineral Resources in South Pacific Offshore Areas (CCOP/SOPAC) has spent a considerable amount of time looking for manganese nodule resources since 1972. Figure 4, Area D, indicates the region that has been surveyed by CCOP/SOPAC. Exon (1983) has summarized all existing data for the region. He used an abundance cut-off of 5 kg/m^2 and a grade cut-off of 2 percent combined Ni + Cu + Co in his analysis of the data. Exon reported findings that tend to confirm those of CNEXO with regard to an inverse correlation between grade and abundance. The only area not showing a strong inverse relationship is in the East Central Pacific Basin. Although the data are extremely patchy, Exon has concluded that this area has economic potential and should be prospected more thoroughly.

A third area of the South Pacific (Fig. 4, Area E) has been of interest because high grade nodules were reported there. The German industrial group, AMR (Arbeitsgemeinschaft Gewinnbare Rohstoffe) conducted reconnaissance surveys in the northern sector of the Peru Basin in 1978 and 1979. The results of these surveys suggest that more exploration is justified on the basis of average values for abundance ranging from 7–14 kg/m^2 and average Ni values ranging from 1.1–1.2 percent. The average copper values, however are less than 10 percent (Halbach, 1984). However, proximity to the South American continent, relatively shallow water depth and high nodule concentrations in some locations may offset the low copper values and warrant further investigation.

North Central Pacific

Cruises were conducted by the Geological Survey of Japan through the years 1974–1978 in the north central Pacific Ocean (Fig. 4, Area F). Compilations of data collected during the cruises show variations within the region in the distribution of manganese nodules (Geological Survey of Japan, 1983).

THE CLARION-CLIPPERTON ZONE

The previous section of this chapter has concentrated on areas outside the northeastern equatorial Pacific in an effort both to present the current level of knowledge and to illustrate the first steps in the process of identifying reserves. The criteria used are rather simplistic and the data are limited by what are publicly available. Most research, however, has been conducted in the primary area of interest: the area between the Clarion and Clipperton Fracture Zones (Fig. 4, Area G). Efforts to assess the resource in this area have been more rigorous since it is assumed that first-generation mining (mining using the first generation of technology) will occur in this area. Presumably, first generation technology will be less efficient and more costly than subsequent models. It follows, therefore, that the first areas of the ocean to be mined will be those with the highest quality and greatest quantity of nodules.

A series of technical reports prepared at Scripps Institution of Oceanography between the years 1976 and 1980 (Hillman and Weldin, 1976; Frazer *et al.*, 1978; Fisk *et al.*, 1979; Frazer and Fisk, 1980) attempted to analyse available data in the Clarion-Clipperton Zone with the objective of determining whether or not there were areas within it that were suitable for first-generation mining. The reports used average Ni plus Cu of 2.27 percent and a cut-off grade of 1.8 percent Ni plus Cu as criteria. Abundance values were reported as averages or broken down into ranges of greater than 10 kg/m^2, 5–10 kg/m^2, and less than 5 kg/m^2. Topography was also considered on a gross scale and frequency distributions were plotted for each element.

Within the Clarion-Clipperton Zone as a whole, just as within the Ocean as a whole, variations in both grade and abundance were evident. Within certain boundaries, however, the Clarion-Clipperton Zone shows less variation in grade than in abundance, and the average grade is significantly higher than the average grade in any other ocean area (United Nations, 1982).

The same areas were examined at one year intervals as new information became available and the assessments were refined or modified based on the new data. A significant aspect of the assessments is that as more information was added, finer distinctions began to show up. Some areas which originally had been looked at as single continuous areas were broken up into two or three exhibiting different characteristics.

The Scripps reports ended in 1980. The same type of analysis was continued and expanded in 1983 by Hillman, however, in an evaluation of three areas in the Clarion-Clipperton Zone. This later study went beyond simple resource evaluation and included estimates of tonnage and grade and descriptions and estimates of the profitability of mining operations in the three areas.

Chapter 2

MINE-SITES: Defining Factors and Size Determination

Once resources have been identified, more intensive investigations are made to attempt to determine the existence of reserves. Additional criteria are applied and sea floor areas are identified that may eventually become mine-sites. It is this process, of identifying manganese nodule reserves, that has been referred to as mine-site definition and the criteria applied as defining factors. This chapter will identify the factors, discuss the values assigned to them and, where possible, try to identify the sources of the values. The factors are generally agreed upon, but the values to be assigned to them are not.

The term "mine-site" is defined here as an ocean bottom area where, under specific geological, technical and economic conditions, a single manganese nodule mining operation can be carried out for a period of time.

There are two principal groups who have provided information and published papers related to the mine-site concept. The first is composed of spokesmen for the mining industry. From that group, information has come from United States based multi-national consortia and the French and Japanese state enterprises. Because each entity has its own disclosure policy, the amounts and kinds of information vary considerably. In the 1970s, the most voluble spokesmen were from Deepsea Ventures, Inc, originally a promoter and now a service contractor for Ocean Mining Associates. Deepsea Ventures presented information through papers given at annual meetings of two professional organizations, the American Mining Congress and the Offshore Technology Conference, and through testimony in the early seventies given in hearings before the United States Congress on various bills related to national deep seabed minerals legislation. Spokesmen from Kennecott Copper Corporation (now KCON) also contributed, but not to the same extent as Deepsea Ventures.

The early papers presented by representatives of the mining industry provide a great deal of valuable information, information which, for the most part, has not been seriously disputed and remains the basis upon which a sizeable body of literature is built.

The second group consists of independent researchers, some of whom are former employees of ocean mining companies and other non-commercial organizations. The papers written by this group focus mainly on estimating the size of the resource and the resulting numbers of mine-sites using many of the values presented by the industry as basic assumptions in the calculations. Many of these papers were prompted by discussions of technical points during the Third United Nations Conference on the Law of the Sea.

In addition, there have been a few papers that have questioned some of the basic assumptions presented by the mining industry; but for the most part, many of the assumptions have been accepted since there is only sparse information available to be used for verification.

FACTORS AND VALUES

In the process of identifying a mine-site, the first set of factors of interest are geological. Starting with a very large area, study of its geology leads to reduction of the area. Adverse macro-topography (mountain chains, seamounts, hills with slopes, etc.) and areas without nodules are eliminated from future prospecting. The remaining area is then examined with economic and technical restrictions kept in mind. Cut-off grade and abundance and adverse micro-topography isolated volcanoes, rock outcrops, faults, slabfields, etc) further reduce the area to that of economic importance.

The quantity of nodules that can be recovered profitably from a given site then depends on locating:

(1) areas within the mine-site where nodules are known to be present;
(2) areas within the mine-site where the grade is above a pre-determined cut-off or where the average grade is above a predetermined value;
(3) areas within the mine-site where the abundance is above a

pre-determined cut-off or where the average abundance is above a predetermined value;

(4) areas within the mine-site where the topography permits the mining equipment to be operated successfully.

The area or areas meeting the criteria listed above is then considered to be mineable.

Economic Factors

Generally speaking, the quantity of manganese nodules extracted from a given mine-site is equal to the product of the annual recovery rate and the duration of the mining operation. The values for these parameters most often found in the literature are three million dry metric tons per year and at least twenty years duration, although up to forty years has been suggested. Hence the amount of manganese nodules that could be brought to the surface from a given mine-site varies from 60 to 120 million dry metric tons.

Duration The original estimate used in the determination of mine-life was apparently linked to incentives in the United States Internal Revenue tax code. In testimony before the US House of Representatives, John Flipse, who was at that time President of Deepsea Ventures, Inc., suggested the reason his company had suggested a 40 year time period.

> I think 40 years was chosen originally because, based upon Internal Revenue Code, financial commitments were going to be written off on pier facilities, buildings, main plant facilities, over a 40-year period. This was the basis for this number being selected for the guarantee. I feel that some of us are ready to accept a lesser period although the drafters and the Mining Congress still feel that the 40 years is a reasonable time period, based upon the durability of a fair proportion of this equipment.
>
> (US Congress, 1974)

As Table 1 shows, the early suggestion of a 40-year duration has given way to a 20-year duration. Pasho (1979) has suggested that the basis for the 20-year duration stems from the investment criteria of financing organizations. He argues that depending on

Table 1. Comparison of suggested annual production and mine life

Source	Annual Production (million dry metric tons)	Mine life (years)
Flipse *et al.* (1973)	1–5	40
Kaufman (1974)	1	20
Siapno (1975)	1	20
Lenoble (1981)	2.1–3	20
OMI-Application (1982)	3	25
OMA-Application (1982)	2.1	20
l'Academie des Sciences (1984)	3	25

when debt financing is obtained in relation to a project's schedule, the types of loans obtained, and the pay-back period, such reserve requirements would easily amount to a time period of about 15–20 years at envisioned recovery rates. Rates of return on the order of 15 percent will be extremely sensitive to durations of less than 20 years, for which they will decrease significantly. At the same time, rates of return are insensitive to extensions beyond 20 years. Pasho suggests, therefore, that for an operation that satisfies industry's rate of return requirements, 20 years duration is adequate to meet the requirements of financing institutions, and would be adequate to allow the maximum practical rate of return for a project.

Recovery Sufficient annual recovery must be established in order to provide adequate feed to a large scale processing facility. The recovery rate, therefore, is probably based on considerations related to the economies of scale of the processing plant. Apart from Deepsea Ventures, the only consortium that has consistently planned to recover manganese in addition to cobalt, nickel and copper, the recovery rate about which consensus has grown is three million dry metric tons per year.

Glasby (1983) has examined this figure simply by working through the arithmetic of a mining operation and has come to the conclusion that "the concept of a three-million-tons-per year-nodule mining operation using a single mining ship is not viable on the basis of the assumptions used. Of course, one could insist on higher nodule densities, a higher ship speed, or a larger dredge head in order to achieve the desired level of operations. Even with maximum numbers fitted into the equation, however, one sees that the three-million-ton-per-year nodule mine-site is likely to be at the limit of technical feasibility, if not beyond it". The same conclusion has been reached by most of the pioneer companies, who envisage the use of two mining ships to produce three million tons per year. The scale effect on the processing facilities is less obvious, suggesting that a 3 million tons per year operation could be less costly per ton of nodule. Therefore, 3 million tons per year for a period of 20 years have been accepted as working values for the purposes of analysis and discussion.

Technological Factors

In order to determine the "minimum" ocean bottom area from which 60 to 120 million dry metric tons of manganese nodules could be recovered, it is necessary to know the efficiency of the mining system, the topography of mine-site bottom as well as the distribution of predetermined cut-offs for nodule abundance and grade.

Other than from limited experiments carried out by industry and research institutions, there is no extended experience with any of the mining methods currently under consideration. In general, the estimates for the efficiency of the mining systems cannot be very accurate unless they are tested over an extended period of time. Further research and experimental work are necessary before any reliable estimates can be obtained for the efficiency of the mining system. On the other hand, regardless of the mining system used, its net overall efficiency depends on two main parameters:

(1) e_d — dredge efficiency, which is the ratio of manganese nodules effectively gathered by the dredge head, versus the nodules lying on the bottom area before it is swept by the dredge head. The range of values found in the literature for dredge efficiency is between 0.3 and 0.85, although some authors consider the range 0.4 to 0.7 as more probable.

(2) e_s — sweep efficiency, which is the percentage of the bottom area actually swept by the dredge before the operator has decided to give up that area. The value estimated ranges between 40 and 75 percent.

The overall efficiency is usually expressed as:

$$E = e_d \times e_s$$

where E = overall efficiency;
e_s = sweep efficiency and
e_d = dredge efficiency

In testimony before the United States House Subcommittee on Oceanography in 1973, Flipse stated:

> Realistic estimates of dredging accessibility and efficiency, sweep efficiency, and the cut off grade of actual minesites indicate an overall mining recovery efficiency of well below 50 percent of the nodules on the minesite.
>
> (US Congress, 1974)

As can be seen from Table 2, not only is the industry still uncertain about definitive values but their published estimates are indeed well below 50 percent. In the Ocean Mining Inc. (OMI) and Ocean Mining Associates (OMA) licence applications to the United States Department of Commerce, collector efficiences are expressed as ranges. Non-industry publications discussing deep sea mining requirements and those estimating the resource have for the most part adopted an overall efficiency value of 20 percent as a standard figure.

Table 2. Comparison of suggested collector efficiencies

Source	Dredge (percent)	Sweep (percent)	Overall (percent)
Flipse *et al.* (1973)	0.30–0.50–0.78	0.45–0.65	0.14–0.16
Kaufman (1974)	0.50	0.65	0.32
Siapno (1975)	0.30–0.70	0.45	0.14–0.32
UN (1979)	0.30–0.80	0.40–0.75	0.12–0.60*
Lenoble (1981)			0.16–0.50
OMI–Application (1982)	0.70–0.80		0.20
OMA–Application (1982)	0.65–0.75	0.40–0.50	0.26–0.38

*Most likely to be between 0.10 and 0.40

Topographic Factors

It is apparent that some areas within the mine-site will not be dredged because the bottom topography is too rough. Areas with tectonic faults, volcanic flows or erosion channels and areas where the bottom sediments are too weak to sustain the mining equipment, as can be the case with recent depositions by bottom currents, will not be mined.

The currently published knowledge of bottom topography and the mechanical properties of bottom soils is certainly not sufficient to obtain accurate estimates for these factors. However, it can be safely assumed that the mine-site will be delineated in such a way as to minimize the unmineable area. It is generally assumed that the unmineable areas will be held below 20 percent of the actual mine-site area.

The proportion of an area that is unmineable due to roughness of terrain is:

$$\frac{A_u}{A_t}$$

where: A_u = area unmineable due to unsuitable topography, and
A_t = total area

As can be seen from Table 3, the early estimates of the unmineable areas within first generation mine-sites were generally lower than later ones.

Table 3. Estimates of mineable portion

Source	Topographic Factor	Proportion below cut-off grade	Proportion below cut-off abundance (kg/m^2)	Mineable Portion
Flipse *et al.* (1973)	0.15–0.20–0.25	0.10		0.75–0.70–0.65
Kaufman (1974)	0.15	0.10	(0.10–0.20)[a]	0.75
Siapno (1975)	0.25	0.10		0.65
UN (1979)				0.60–0.90[b]
Lenoble (1981)	0.20		0.12–0.25	0.68–0.55
OMI–Application (1982)	0.37	0.15		0.48
OMA–Application (1982)	0.30–0.45	0.10–0.15	0.10–0.15	0.50–0.25

a Not used in formula.
b Agreed upon range; no formula provided.

Apparently early assumptions about the smoothness of the bottom were modified as more intensive exploration took place. An example is given by one author of an area that had been surveyed with a 5 statute mile vertical and a 10 statute mile horizontal grid. It was assessed as “deep sea plain” but later turned out to be mountainous and consequently partly inaccessible to mining technology (Amann, 1975). Furthermore, "large smooth portions of the sea floor rarely contain any manganese nodules. This situation results because smooth, flat areas of the sea floor are associated with high rates of sediment deposition. A high rate of sedimentation precludes the formation of manganese nodules or if present, would be buried" (Kaufman, 1974). Finally, "occurrences of crusts, scarps, cliffs, and 'pot-holes' from one hundred metres to several kilometers in diameter with depths of more than 50 metres with steep edges, have been reported. For instance cliffs, related to tectonic faults are edging the flanks of most of the north–south hills" (Le Suavé, 1985).

Nodule Distribution Nodules are known to be present throughout the World Ocean (Fig. 1) as well as in some freshwater lakes. However, the presence of nodules in large quantities neither indicates the amount of *recoverable* resources nor guarantees the economic viability of a mining operation. There has been a tendency to assume that the mere occurrence of nodules is synonymous with a mine-site. This has led to unrealistic expectations about the magnitude of the recoverable resources.

Figure 5 illustrates a hypothetical "realistic" mine-site showing patches of mineable area. Halbach and Fellerer (1980) state that "it must be kept in mind that the so-called mine-sites consist of more or less extended single "spots" which are separated by zones with low-grade nodules or absolutely sterile regions, by mountain chains, seamounts and volcanoes or by other hazards. Due to technical reasons, each such spot must cover not less than 500 km^2, in order to be mineable". The technical reasons for this

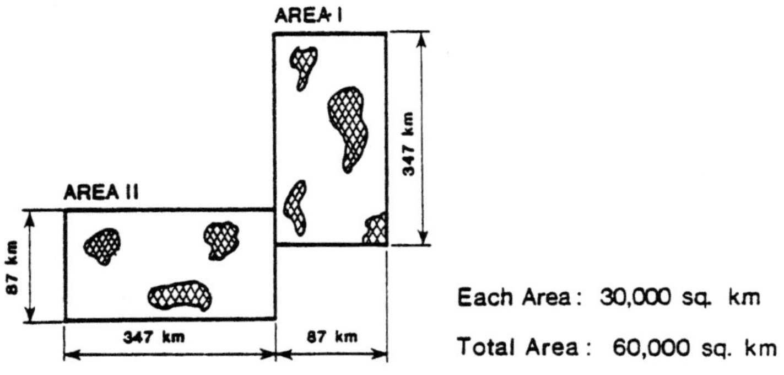

Fig. 5. Hypothetical "Realistic" Mine Site from: Andrews *et al.*, 1983 (Fig. 22).

limitation are not amplified but are presumably related to the technology to be used. In an earlier paper by Fellerer (1975), it is reported that no continuous deposits larger than about 10 km^2 with uniform genesis and population density appear to exist.

Nodule Grade and Abundance

In addition to unacceptable topography and barren areas, other areas within the mine-site will be rejected for mining because the average nodule abundance is not sufficiently high to allow normal production rates to be maintained or because the average grade of the available nodules is too low to be acceptable for processing.

Published data on abundance and grade does not allow for accurate estimates of areas falling below cut-off values. However, it appears that the industry is more concerned with the cut-off values for abundance than for grade. The principal argument appears to be that if sufficiently high average grade is found in the prime areas, with small variations over short distances, then discrimination between the cut-off grade areas and the rest of the mineable area cannot be effectively achieved. On the other hand, the cut-off value for abundance is considered critical. According to current knowledge, even in the best areas, nodule abundance varies considerably over short distances, sometimes even over distances less than 1 km. The nodule field appears as a blanket with a large number of holes where the manganese nodules cannot be found (Lenoble, 1980).

If the size and density of the voids vary considerably, in some cases they may reduce the average abundance below the cut-off value. It cannot be assumed that all hollow areas will be detected effectively with sufficient accuracy and avoided. Therefore, it is estimated that even in the best mine-sites, hollow areas which can be avoided by the mining device will constitute at least 10 to 20 percent of the total mine-site.

Grade The cut-off grades for four metals suggested by Kaufman (1974) were:

Mn ....20.0 percent
Ni................1.0 percent
Cu ....0.8 percent
Co ....0.2 percent

It is important to remember, as Kaufman pointed out, that

these were based on one "viable economic model", presumably, a Deepsea Ventures model, and that as such they are linked to the mining system, processing methods, distribution costs and metal prices projected by Deepsea.

CNEXO has recorded that "metal content must also exceed a certain limit determined by the metals to be extracted, treatment efficiency, production costs, and metal prices. The economics of the production process remain controversial". (Bastien-Thiry, *et al.*, 1977). In a paper published in 1980, Lenoble suggests minimum average grades. He lists characteristics of two mine-sites of different sizes. In one example, he says "the average grades of the nodules are taken to be higher than 27 percent Mn, 1.35 percent Ni, 1.15 percent Cu, 0.2 percent Co ..." and in a second model, "the metal content of the nodules is likely to be at least 1.4 percent Ni, 1.2 percent Cu, 0.25 percent Co".

As with so many of the factors that define an ocean mine-site, the Deepsea Ventures suggested value of 1.8 percent (Cu+Ni) generally prevailed as the accepted cut-off value for average grade. However, some years later, it was foreseen that copper and nickel alone would not provide a sufficient return; manganese and cobalt were then considered as important as nickel and copper. The concept of "metal equivalent" was then introduced. Metal equivalent is expressed as the grade of a reference metal, in this case, nickel, that would produce the same revenue as the sum of the revenues produced by all the metals recovered from manganese nodules. The ratio of the expected price of each metal to that of the reference metal and the recovery ratio of each metal must therefore either be known or inferred. For instance, as shown in Table 4, the nickel equivalent of nodules that contain 1.3 percent nickel, 1.2 percent copper, 0.25 percent cobalt and 29.0 percent manganese with an expected recovery of 95 percent for nickel, 80 percent for copper, 80 percent for cobalt, 85 percent for manganese and price ratios versus a nickel price of 0.3 for copper, 3.0 for cobalt and 0.07 for manganese is 3.76 percent.

As a result, cut-off nickel equivalents can be used to delineate areas which contain nodules above a predetermined value, with a corresponding average nickel equivalent that is obviously above the cut-off value.

Table 4. Nickel Equivalent

Metal	Grade percent	Price ratio to nickel	Metal Recovery	Nickel Equivalent
Nickel (Ni)	1.3	1.0	0.95	1.24
Copper (Cu)	1.2	0.3	0.80	0.29
Cobalt (Co)	0.25	3.0	0.80	0.60
Manganese (Mn)	29.0	0.07	0.85	1.73
Ni Equivalent				3.76

Abundance Estimating nodule abundance is one of the most problematical areas in defining an ocean mine-site. Estimates of average abundance values that have a solid technical basis have been notably lacking. In part, this is due to the difficulty of obtaining reliable data. Efforts have been made by the mining industry to improve the reliability of abundance values but these efforts have been only partially successful. For instance, Bastien-Thiry (1979) states that "despite all the precautions and all the care taken, in the final analysis the nodule concentration figure obtained may be as much as 50 percent off". Furthermore abundance values in the public domain have not been very helpful due to unreliable and sparse data (United Nations, 1982).

The cut-off abundance value which has tended to prevail in the literature is 5.0 kg/m^2. The original source of the figure again seems to be Deepsea Ventures. "With presumably envisioned mining systems, the lowest concentration that can be feasibly mined is about 1.0 lb/ft^2 (4.88 kg/m^2). In a typical mine-site, therefore, about 10–20 percent of the deposit may contain concentrations too low to be mined" (Kaufman, 1974).

CNEXO has not committed itself to a single cut-off figure but has discussed the importance of the cut-off value and has used a cut-off value of 9 kg/m^2 to estimate the amount of area needed to support a mining operation (Lenoble, 1980).

Estimates of the percentage of area to be eliminated due to abundance falling below cut-off have been given by only a few authors from the commercial sector (Table 3). It is interesting to note that in the early papers by the mining industry, cut-off abundance is discussed but is not used in the calculations of mine-site size. In the estimates of mineable area, therefore, a cut-off abundance proportion is sometimes included and sometimes not.

Another method that has been utilized in taking into

account the reliability of abundance values again avails itself of the concept of metal equivalence. By combining grade and abundance, the quantity of metal contained in nodules per unit surface area can be calculated. The grades of nodules are expressed in dry weight while abundance is expressed in wet weight per unit surface area. The reduction factor used to arrive at the former is 0.7. The set of metal grades can be replaced by a metal equivalent as defined earlier, giving a quantity of metal equivalent per unit surface area. By multiplying the quantity of metal by a given area, the tonnage of *in-situ* metal or recoverable metal equivalent in the area can be computed.

Mineable Area

The factors that limit the area that is mineable can be combined as follows:

$$M = A_t - (A_u + A_g + A_a)$$

where:
M = mineable area
A_u = area unmineable due to topography
A_g = area below cut-off grade
A_a = area below cut-off abundance
A_t = total area

Using the above formula and published values, Table 3 shows the ratio of mineable portions in the most prospected areas by industry following 10 to 15 years of work. The biggest change has come in the proportion of area considered unmineable due to topography. The adjustment seems to reinforce the idea that initial assumptions about the nature of the seabed were optimistic.

As work progressed during prospecting in the estimation of mineable areas within a potential mine site, it became increasingly clear that a considerable amount of data has to be acquired to accurately map the topography of the site, the occurrence of obstacles within it, nodule abundance and nodule grade. The extent of the work to be done during exploration to reduce the present margins of error in the estimation of each of the four factors above is briefly summarized:

(i) *Topography*: Since the hills and volcanoes in the most

prospected and most promising areas are several kilometers wide, accurate maps depicting the topography of potential sites should be at a scale of 1:50,000 or above. By using multi-channel echosounders, the topography can be surveyed with a 3 kilometer spacing between adjacent profiles. At a speed of 10 knots for example, an area of 10,000 km^2 would require between eight and ten days to be accurately mapped.
(ii) *Obstacles*: The presence of scarps, cliffs and large pot holes in potential sites cannot be detected by surface equipment and requires the use of a high resolution side-scan sonar towed at about one hundred meters above the ocean bottom. Using such equipment, a coverage of about one kilometer per tract can be achieved at ship speeds of up to 1.5 knots. An area of 10,000 km^2 would therefore require about 9 months to be completely and accurately mapped. During the prospecting phase of manganese nodule development, only a few companies or institutions used such equipment and, even then, only for relatively short periods of time. Most of the data on the occurrence of obstacles within potential sites are, therefore, very patchy having been acquired from irregular and scattered tracks. As a result, the importance of obstacles in the determination of mineable areas has either been underrated by many researchers or only statistically inferred by others. During exploration, to undertake the obstacle survey using the sonar described above, the likely strategy would be first statistical sampling (inceasing the spacing between profiles to 5 km or a 20 percent coverage) for area elimination and mining requirement definition and then undertaking systematic exploration in advance of mining by attaching a front-looking sonar to the mining head.
(iii) *Nodule abundance*: Most of the data acquired on abundance were obtained using optical systems: still movie and TV cameras. As a result of their limitations, such as their areal coverage and the time required to survey large areas, the information gathered on abundance during prospecting was not continuous but rather scattered. New equipment, probably acoustic systems towed near the seabed, have to be developed. The strategy used during exploration to determine abundance in the mine site to an acceptable level of confidence will be similar to that used in the obstacles survey; essentially, statistically "sampling" for area elimination and mining requirement definition, then systematic pre-mining exploration during the commercial stage. It must be noted that the delimitation of first geneation mine sites should be

such that the discarded areas should be mineable when new and more efficient systems become available.

(iv) *Metal content*: The only method available at the present time for obtaining data on metal content is through sampling and then chemical analysis. It is unlikely that other means, such as bottom radiometric measurements, for example, can be developed to guarantee a fast and accurate survey, of this parameter. However, as the prime areas within which the potential mine sites are to be found were selected on the basis of the apparent continuity of high metal grades over large areas, non-profitability of a mine site due to low grade appears highly unlikely. Nevertheless, sampling must be done to verify this continuity. During exploration, such sampling will be limited to areas selected on the basis of the previous parameters: topography, obstacles and abundance.

From the above considerations it follows that the estimates for the mineable area within a mine-site vary between 50 and 90 percent, the mineable area being the total mine-site area minus the sum of the unmineable areas and those areas within which grade and abundance fall below the required cut-off values.

SIZE

Another subject for discussion among researchers has centred around the problem of estimating the size of individual mine-sites. Although there is agreement among the different authors with respect to the basic formula used to calculate the size, and the factors affecting the size, of single mine-sites, the estimates vary within a very wide range, by an order of magnitude. The main reasons for such large variations seem to be: lack of sufficient data in the public domain for use in estimation procedures and methods; lack of actual experience and reliance on limited experiments; differences in the definition for the various factors; and the different points of view taken by scientists, industry and potential regulating agencies.

Sizing Formula

Formula to approximate the size of a mine-site area have been published since 1973 (Flipse *et al.*, 1973; Kaufman, 1974; Siapno,

1975; Pasho, 1979; Lenoble, 1980; Halbach and Fellerer, 1980; Halbach, 1984; Ocean Minerals Company, 1982; Ocean Management, Inc., 1982). All the formulae are basically the same although there are variations in some of the parameters used.* Determination of mine-site size can be generally expressed by the following formula:

$$A_s = \frac{(A_r)(D)}{(C)(E)(M)}$$

where A_s = size of the mine-site km^2
A_r = annual nodule recovery rate (dmt/yr)
D = duration of the mining operation (years)
C = average nodule abundance in the mineable areas (dmt/km^2)
E = overall mining efficiency (percent)
M = proportion of area mineable

When this formula has been applied, it has often been used to illustrate a "typical" mine-site or a range of sizes which indicate the probable lower and upper limits that could be expected. Table 4 summarizes the sizes calculated by using the above formula or a slight variation of it.

The range of values reported in the literature varies between about 12,000 and 130,000 km^2, but the prevailing number of estimates is between 35,000 and 60,000 km^2. The wide range in the estimates is primarily due to the different perceptions of the risk factors involved.

Table 6 shows the factors used for two hypothetical cases – "best" and "worst" – and the calculated size of the mine-site area. As can be seen, the range between the two values is almost an order of magnitude. In reality, some combination of parameter values will appear and hence the required area for a mine-site will be specific for every case.

Thus, the area needed to sustain a single mining operation, that is a mine-site, will be determined by a combination of the limits to recoverability imposed by the nature of the nodule

*Most variations in the formula occur in the parameters used to calculate "*M*". Some authors use a relationship where the areas considered unmineable are summed and subtracted from 1, while others calculate a product of ratios of mineable areas.

deposits, the topography, the overall mining efficiency and the total quantity of nodules that must be recovered from the area to ensure commercial viability.

Table 5. Ranges and estimates of mine-site size

Source	Size (km^2)
Flipse, *et al.* (1973)	32,668–127,043
Kaufman (1974)	11,500
Siapno (1975)	23,000– 54,000
Pasho (1979)	38,000
Lenoble (1980, 1981)	17,500– 94,000
Halbach (1984); Halbach and Fellerer (1980)	80,000–120,000

Table 6. Hypothetical cases for sizing a mine-site

Parameter	Values for		Units
	Best	Worst	
1. annual recovery	3	3	million dry metric tons
2. duration	20	40	years
3. average abundance	10	10	dry kg/m^2
4. sweep efficiency	0.7	0.4	percent
5. dredge efficiency	0.7	0.4	percent
6. unmineable factor 1	20	10	percent
7. unmineable factor 2	10	20	percent
Area size	15,300	125,000	Km^2

Chapter 3

ESTIMATES OF AVAILABLE MINE-SITES

In 1975, at US Congressional hearings on a seabed hard minerals bill, Marne Dubs, Director of the Ocean Resources Department of Kennecott Copper Corp., said that some 200 potentially commercial mine-sites had been identified (US Congress, 1975). Again in 1979, Kennecott reiterated that 200 mine-sites existed (Halkyard, 1979). In neither case, was supporting information provided nor was any distinction made between first or second generation sites. Apart from general statements such as these, the US mining consortia have offered little additional information concerning their estimates of available mine-sites.

Within the mining industry as a whole, the CNEXO group is the only one to have published mine-site estimates and an outline of the basis for their estimates. Given that miners are interested in the discovery and development of one or two actual mine-sites, it is not surprising that they do not appear to have addressed the question of the total number of mine-sites that might ultimately be available.

This question has, however, been taken up by other researchers. Over the last 10 years, about a dozen papers have been published that attempt to estimate the number of available mine-sites. The following discussion reviews the most important estimates.

The general approach has been to calculate the approximate amount of recoverable nodules and then divide that total tonnage by a factor reflecting projected annual recovery, recovery efficiency and mine life. This number of sites therefore equates to mining operations. Some have projected world-wide estimates, but most have limited their consideration to estimates within the Clarion-Clipperton Zone.

Other than giving a general indication of the geographic areas for which the estimate is being made, the discussions do not deal with specific sites. Before reviewing the individual estimates, the reader should be aware of some of the general problems that make any evaluation of the estimates difficult.

(1) "Prime area" is a term used to designate seabed areas that appear to have the highest relative potential. The degree of uncertainty accepted by each author varies widely. This is particularly true for global estimates where there is far less information than in the Clarion-Clipperton Zone for example, and must therefore depend to a large extent on geological inference. Other factors being equal, (e.g. percentage of area above cut-off, percentage of area unmineable due to topography), the larger the estimate of prime area, the greater the number of mine-sites.

(2) A second topic of controversy is the relationship between grade and abundance. In the simplest terms, if grade and abundance are treated independently but are, in fact, not independent, then an overestimation of the number of mine-sites could result. While the mining industry has maintained that they are independent variables, their assumption was challenged in 1978 by Menard and Frazer who suggested an inverse correlation between grade and abundance. Although others have confirmed this finding, McKelvey *et al.* (1979) showed a correlation coefficient between nickle plus copper content and abundance of −0.24 at the 98 percent confidence level and concluded that for the Clarion-Clipperton Zone, the data were "weak enough to suggest that *in this area* only a slight positive bias would result from treating grade and abundance as independent variables ..." [emphasis added]. It can be concluded, therefore, that although grade and abundance may be inversely correlated, serious errors in the number of mine-sites should not result even if they are treated as independent variables.

(3) A third issue that affects estimates of the number of mine-sites is the limited quantity of data that is publicly available. This topic has been discussed thoroughly in Volume I of this series (United Nations, 1982). Apart from estimates based on CNEXO data and a recent estimate based on both public and proprietary AMR data, all other estimates have had to rely solely on information in the public domain.

(4) In addition to different assumptions and methods used in

the estimation of prime areas, different assumptions are also made in the recovery efficiencies, annual nodule recovery and the mine life. Recovery efficiencies are usually either 20 or 25 percent; annual recovery varies from 1 million tonnes to 4.5 million tonnes which is at times on a dry basis and at other times on a wet basis; and mine life is either 20 or 25 years. While this set of assumptions can be the cause of differences in the number of mine-sites estimated, they are not as problematic as differences in the assumptions used to estimate prime area. Calculation of prime area depends on knowledge of the seafloor and the resource, combined with cut-off factors derived from production requirements. The methods used cannot be adjusted to a common basis, whereas the system factors and production requirements listed above can be easily adjusted to a common basis.

Global Estimates

Three researchers have made serious attempts to estimate the number of mine-sites worldwide: Archer (1976, 1979, 1981), Holser (1976) and Frazer (1977). A full discussion of estimation methods can be found in Volume 1 of this Series (United Nations, 1982).

Archer (1976) used two methods to estimate the amount of prime area available for first generation mining. Using the so called Global Estimator, he calculated an area of 3.0 million km^2, and using the Prime Area Estimator, he arrived at an area of 1.3 million km^2. Archer clearly recognized the limitations of the two methods and expressed a preference for the latter. Assuming 20 and 25 percent overall recovery efficiencies, 3 million (dry) tonnes per year recovery rate and mine lives of 20 and 25 years, Archer estimated the maximum number of first generation mine-sites to be between 106 and 166. Using the Prime Area Estimator alone, with efficiencies of 20 and 25 percent and mine lives of 25 and 20 years, his estimates range from 44 to 69 mine-sites. In 1979, Archer revised his estimate of total first-generation mine-site area to "very approximately" 2.2 million km^2. He also revised his minimum required abundance value downward from 10.0 kg/m^2 to 5 kg/m^2. Using the same sets of assumptions for efficiencies and mine life as in 1976, the resulting estimate is 60–95. Archer's latest estimate, in 1981, resulted in a range of 30 to 75 three million (dry) tonnes a year operations.

Holser (1976), in addition to revising Archer's early estimates, suggested that there might be between 80 and 185 first generation mine-sites. His estimate was of prime area 9.9 million km^2 using the Global Estimator and 4.2 million km^2 using the Prime Area Estimator. Given the much larger amount of prime area, Holser's estimates would have been much greater if he had used the same minimum required abundance as Archer (10.0 kg/m^2). However, he used an average abundance of 10.0 kg/m^2 citing unpublished information from mining companies. Holser also estimated the number of third generation mine-sites to be between 190 and 460 assuming a recovery efficiency of 50 percent. Holser states that the Prime Area Estimator approach is a conservative estimate and that the Global Estimator approach is more representative of the probable potential.

Frazer (1977) took a more conservative approach than either Archer or Holser. She used a Grid Estimator method that limited prime areas to 5°-squares where data was publicly available. Her final assessment of prime area was 1.37 million km^2 and her estimate of the possible number of mine-sites was between 14 and 56. Frazer, like Archer (1979, 1981), used a 5 kg/m^2 abundance cut-off.

Frazer offers several reasons for the disparities between her estimates and those of Archer and Holser. Archer derived an estimate of the prime area (1.3 million km^2) similar to that of Frazer, but Archer used an average abundance of 18.3 kg/m^2

Table 7. Estimates of the number of first-generation mine-sites worldwide

Source	*Total Area (millions km^2)*	*Number of mine-sites*
Archer (1976)	1.3*	44–69
	3.0	106–166
Archer (1979)	2.2*	60–95
	4.9	140–215
Archer (1981)	2.2	30–75
Holser (1976)	4.2	80
	9.9	185
Frazer (1976)	1.37	14–56

*Preferred estimate

while Frazer used an average of 11.0 kg/m^2. Holser estimated that 54 percent of the prime areas would have abundances greater than 10 kg/m^2, while Frazer estimated that only 38 percent would be greater than 5 kg/m^2. Finally, Frazer's method was far more restrictive in the selection of prime areas since it was limited to areas where data were publicly available.

The principal estimates of available mine-sites worldwide are summarized in Table 7.

Clarion-Clipperton Zone Estimates*

In any discussion of manganese nodule resources, the Clarion-Clipperton Zone deserves special consideration. Not only has most exploration taken place in this area and most of the scientific research on manganese nodules been conducted there, but geologically, the area appears to be unique in the quantity, quality and extent of the resource. While estimates of the number of mine-sites in the Clarion-Clipperton Zone exhibit a narrower range of values than those for the World Ocean, Frazer (1980) attempted to reconcile the apparent discrepancies by converting each author's results to a standard basis. The outcome of defining tonnage of economically mineable nodules as having grade greater than or equal to 1.8 percent (Ni+Cu) with an average grade of 2.25 percent and abundance greater than, or equal to, 5 kg/m^2 is reproduced in Table 8.

Frazer concluded that demonstrated paramarginal resources in the Clarion-Clipperton Zone were estimated to be from 4 to 15 billion tonnes of nodules and that differences among the authors she compared were not as great as had been previously assumed. At 24 percent mining efficiency and production requirement of 75 million dry tonnes per mine-site, the region would provide enough nodules to sustain 9 to 34 mine-sites.

Archer (1976) predicted between 23 and 37 mine-sites in the same area based on the same assumptions that he used in his prediction of sites available worldwide. He warns, however, that it is probable that there will be fewer discrete sites than what is theoretically possible since the estimate does not account for areas unmineable due to topography, and areas that may be too small and too widely dispersed to make mining practicable.

---

*The Clarion-Clipperton Zone was defined broadly to be between 0° and 30°N and 170°E and 100°W.

Table 8. Demonstrated paramarginal resources in the Clarion-Clipperton Zone (Amounts *in situ* in millions of tonnes) a/

Source	*Nodules*	*Nickel*	*Copper*	*Cobalt*	*Manganese*
Archer (1975)	12,400	106	87	87	2,400
Holser (1976)	14,000	120	98	24	2,700
Pasho and McIntosh	11,900	38	79		
(1976)	(1,600–50,600)	(14–352)	(9–324)		
Pasho (1977)	(3,600–34,200)				
Frazer (1977)	13,800	121	100	23	2,400
	(6,900–27,600)				
Mero (1977)	54,000	650	520	115	11,000
Bastien-Thiry	4,750	41	34		
et. al. (1977)	(3,900–5,600)				
McKelvery *et. al.* (1978)	15,000	135	104	23	2,600
Frazer (1980)	(4,000–15,000)				

a/ Each author's results were converted to the common basis of *in situ* tonnes.
From: Frazer (1980), Table 2, p.58.

Holser (1976) provided a figure which shows that the Clarion-Clipperton Zone will yield only about 20–25 first-generation mine-sites. It has been pointed out that since this number is about one quarter of the total number of first-generation mine-sites predicted by Holser, the importance of other regions relative to the Clarion-Clipperton Zone seems to be overstated (Charles River Associates, 1982). It is interesting to note, however, that while Holser's worldwide predictions are significantly higher than those of other researchers, his Clarion-Clipperton Zone estimates are in substantial agreement with the majority.

Pasho has written two papers estimating the number of mine-sites in the Clarion-Clipperton Zone. The first paper (Pasho and McIntosh, 1976) estimated from 5 to 140 mine-sites at the 90 percent confidence interval. The second paper (Pasho, 1977) refined that estimate to a narrower range of from 10 to 95 mine-sites. In both papers, a Monte-Carlo statistical technique was used to combine the estimated probability distributions of values of all the individual factors to arrive at a probability distribution of the quantity of recoverable nodules.* The median estimate for the Clarion-Clipperton Zone was 27 mine-sites.

Frazer (1977) using the Grid Estimator method, predicted about 26 mine-sites in the Clarion-Clipperton Zone. Frazer stressed two points: (1) the estimate is based only on 5°-squares that contain data meeting her criteria, and (2) the method does not account for seafloor topography. The first point could have the effect of making her estimate too low and the second point could have the effect of making her estimate too high. However, additional data does not necessarily mean additional resources. The Grid Estimator method was applied to the South Pacific region in 1982 after the data base in that area had increased almost twofold. Using the same criteria as Frazer, it was found that with an increase in data, the one 5°-square that previously met the criteria no longer did so and no other squares were found (Fisk, 1982). Recognizing some of the shortcomings of her method, Frazer felt that her estimate could be off by about 50

* The major source of estimates for each factor was the subjective probability distributions provided by each consortium. This was in essence a delphi approach to establishing the distributions that were then combined using Monte Carlo techniques. Pasho established his own distributions for areas with adequate grade based on published analyses and IROR-DCF models for various operations.

percent in either direction and she cited a range of 14–56 first generation mine-sites.

Bastien-Thiry and Lenoble have published two papers presenting estimates of numbers of mine-sites. The estimates are based on data collected by CNEXO. While other researchers refer at times to industry contributions to their estimates, their primary source of information and data was the public data base. The French work is based on an entirely different data base and their results are significantly different. In the earliest paper (Bastien-Thiry, *et al.*, 1977), the estimated number of mine-sites for the Clarion-Clipperton Zone was between 8 and 11. The later paper (Lenoble, 1980) estimated between 3 and 10 mine-sites. Frazer (1980), in analyzing the French results, found that the main differences between the CNEXO data set and the public data set that she and others had used were in the estimates of nodule abundance. The French estimates were lower in both the percentage of area above the cut-off of 5 kg/m^2 and the average abundance within that area. The French results challenge a basic assumption about the relationship between population and abundance as proposed by Kaufman (1974) and, subsequently adopted by Archer (1976), Holser (1976) and Frazer (1976). Kaufman proposed that a 35 percent population (surface of the ocean bottom covered by nodules) would correspond to an abundance of 9.76 kg/m^2 based on an average nodule diameter of 3.71 cm and a flatness of 0.57. The French contend that the *actual* data were: diameter, 4.0 cm; flatness, 0.56; roundness, 0.73 and population, 14 percent, giving an average nodule abundance of 3.51 kg/m^2 (Lenoble, 1980). Without additional information, the estimates cannot be reconciled.

McKelvey, *et al.* (1979), using an updated public data base, predicted about 27 mining operations in the Clarion-Clipperton Zone. They estimated the area to be about 2.5 million km^2 where nodules contain more than 1.8 percent nickle plus copper. They then estimated that half that area contains nodules above the 5 kg/m^2 abundance cut-off. They found the average abundance in that area to be 11.9 kg/m^2. They further assumed a 20 percent recovery efficiency and estimated the potential recoverable resources to be 2.1 billion dry tonnes of nodules.

Halbach and Fellerer (1980) and Halbach (1984) have predicted a range of 14–30 first generation mine-sites in the Clarion-Clipperton Zone in a prime area estimated at 0.6 km×106 km. In addition, they estimate 7–15 second-generation sites for the same area and 45–100 first-generation sites for the whole Pacific Ocean. The estimate of the prime area is based on the results of many years of field experience and the data of numerous exploration cruises (Halbach and Fellerer, 1980). The authors use a cut-off abundance of 8 kg/m^2, (slightly higher than values used by other researchers), a grade cut-off as used by previous researchers and, unlike most other authors, take account of topography by estimating that 60 percent of the area will be unmineable due to obstacles of various kinds. However, the initial estimate of the size of the zone was 6 million km^2, so the size of the area after elimination of topographically adverse areas is 2.4 million km^2, a size comparable to that estimated by others. A second difference in values is the recovery efficiency which is estimated to be 35 percent. The end result is that the number of sites they project for the zone is within the range of previous estimates.

Estimates from the papers discussed above are summarized in Table 9. The consensus of opinion seems to indicate that there could be about 25 mine-sites in the Clarion-Clipperton region with about 50 percent error in either direction although the lower bound of the estimate by Pasho (1977) and Halbach and Fellerer (1980) allow for the possibility of the CNEXO estimate.

Table 9. Estimated number of first-generation mine-sites in the Clarion-Clipperton Zone

Author	*No. of Mine-sites*	*Duration of Operation (years)*	*Tonnage per Site (millions tonnes)*
Archer (1976)	23–37	25	75
Holser (1976)	20	lifetime*	75
Pasho (1977)	10–95	20	75
Frazer (1977)	26	25	75
Bastien-Thiry (1977)	8–11	20	75–80
McKelvey (1979)	27	lifetime*	75
Halbach and Fellerer (1980)	14–30	25	75

*probably 20–25 years but unspecified by the author.

Chapter 4

EXPLORATION AREAS

The concept of an exploration area, from a regulatory or legal point of view, is that of an area that can be held and explored exclusively by someone who has acquired a legal right to do so. The question of how large an exploration area for a future manganese nodule mining operation should be, is of concern not only to potential miners, but also to those who write laws and regulations governing seabed mining. In estimating the total area required for exploration, the potential nodule miner's primary goals are to (i) confirm that the candidate deposit is in fact an economic deposit and (ii) to acquire the specific information pertinent to the confirmed ore body for the purposes of equipment design for mining and processing. In respect of confirming that the candidate deposit is an economic deposit, the nodule miner's objective is to locate as many mineable areas within a potential mine site as he can. Since nodule grade and nodule abundance vary between oceans and within an ocean area, and the sea floor surface, as defined by the slopes and obstructions on it, is critical in the design of minimum technology, it is these considerations which determine whether patches within a potential mine site are even mineable.

Over the claim site, and in the mineable areas, the *in situ* nodule resources must be of sufficient tonnage to support an economically viable project. The mineable tonnage must be greater than the required total production over the life of the mine since it will not be possible to mine the area at 100 percent efficiency. Up to the present time, the two organizations that have been most involved in this process from a regulatory point of view are the National Oceanographic and Atmospheric Administration (NOAA) of the United States Department of Commerce, and the United Nations.

This chapter examines the relationship between the size of an exploration area and the size of a mine-site and describes the present status of requirements for exploration area sizes as reflected in the United Nations Convention on the Law of the Sea and the regulations associated with United States Public Law No 96–283, the Deep Seabed Hard Mineral Resources Act.

RELATIONSHIP BETWEEN EXPLORATION AREA AND MINE-SITE SIZE

Information on the relationship between the size of the exploration area and the size of the area where mining will take place is scarce. In 1975, Amann stated that "there should be no rules as to definite sizes of exploration titles. The areas must be large, however, on the order of 100,000 km^2 and should be subsequently reduced by factors of 5 to 10 percent to arrive at exclusive mining tracts of suitable sizes". In 1980, during the Third United Nations Law Conference on the Law of the Sea, the question of exploration area size became an issue as the participants sought to define a "pioneer area"* in drafting provisions concerning preparatory investment protection, provisions which eventually became Resolution II. The negotiation eventually resulted in a pioneer area of 150,000 km^2 which, after eight years, would be reduced by half. Thus a minimum ratio of 2:1 was considered acceptable by the parties involved in the negotiations. In 1982, OMCO, in its application for an exploration licence to NOAA, stated the following: "Based on the level of survey work over most of our areas of interest and the confidence associated with these results, we estimate that the exploration licence area must be

---

*As defined in Resolution II, para. 1(e), a "pioneer area means an area allocated by the Commission to a pioneer investor for pioneer activities pursuant to this resolution". "Pioneer activities" are defined in para. 1(b) as:

> ... undertakings, commitments of financial and other assets, investigations, findings, research, engineering development and other activities relevant to the identification, discovery, and systematic analysis and evaluation of polymetallic nodules and to the determination of the technical and economic feasibility of exploration. Pioneer activities include:
>
> (i) any at-sea observation and evaluation activity which has as its objective the establishment and documentation of the nature, shape, concentration, location and grade of polymetallic nodules and of the environmental, technical and other appropriate factors which must be taken into account before exploitation;
>
> (ii) the recovery from the Area of polymetallic nodules with a view to the designing, fabricating and testing of equipment which is intended to be used in the exploitation of polymetallic nodules. . . .

several times the size of the permit area.* This ratio may vary with geographical location and is also dependent upon the density of survey data employed in the licence area selection".

ESTIMATION OF EXPLORATION AREA SIZE

One of the long standing problems in the literature on mine-sites is the use of terms. This is particularly evident when looking at the terms used for "mine-site". As has been shown in Chapter 2, the mathematical formula developed to determine the size of a mining area was clearly meant to define a mine-site. However, the formula has been used to calculate areas called "tracts", "mine-sites" and more recently, "exploration areas". In other words, approximately the same formula is used to calculate an area of 11,500 km^2 called a "required tract" (Kaufman, 1974), and an area of 136,000 km^2, called an "exploration area" (Ocean Management Inc., 1982). It has been shown that changes in the variable values can result in wide variations in the size estimates. Given the ease with which variations in size can be introduced and the uncertainties associated with many of the values used, the purpose for which the calculation is being used must be taken into consideration.

During the Law of the Sea negotiations, different points of view were expressed leading to two different suggestions.† One suggestion was that the size of a "pioneer area" be approximately 60,000 km^2; another was that the maximum size be set at 150,000 km^2. The rationales expressed were different. The argument for 60,000 km^2 was made as follows:

(1) In view of the fact that potentially mineable areas for first generation mining are limited in size, the number and size of the individual pioneer areas allocated to applicants should be limited to the extent necessary for encouraging the continued activities of the entities which have been engaged in exploration for deep seabed mineral deposits as well as for reserving maximum mineable areas for future activities of new-comers entering into deep seabed mining.

---

*US law grants licences for exploration and permits for mining.

†The following discussion is based upon information provided by the United Nations Secretariat.

(2) In order to avoid conflicts arising from applications for overlapping areas, it is necessary to limit the size of a pioneer area.

The rationale for 150,000 km^2 emphasized the level of risk involved and its relationship to area size. It was argued that an applicant would certainly face economic disaster if he discovered after detailed exploration that the area granted by the licence did not contain such tonnage of mineable ore absolutely necessary for an economically viable project. A zero risk would require an area of 210,000 km^2. Since it would be hard to conceive that a private investor could be compelled to accept a risk of economic failure of more than 10 to 20 percent, the minimum required size of a licence area would be approximately 135,000 to 155,000 km^2. The technical support for this exploration area size was established by using a risk analysis technique with global mean values reflecting the average of thousands of samples from the Clarion-Clipperton Zone. Since global mean values are invariably lower than mean values for mineable area, the area size calculated will invariably be greater and will give the potential miner a greater assurance of success. This approach was apparently convincing in the Law of the Sea Conference and resulted in setting a maximum size limit of 150,000 km^2 for a pioneer area.

A similar approach seems to have been used in some of the calculations contained in the licence applications to NOAA, although the language used is somewhat ambiguous at times. OMI states that "high percentages of concentration and uniformity of population in areas exceeding several square kilometers have been found to be rare. Similarly, areas with zero coverage have been identified separately and have been eliminated from such assessments of concentration" (Ocean Management Inc., 1982). Although obvious holes are eliminated, the mean value which will eventually be used to compute the areal extent of the mine-site itself is not the value being used in the calculation of exploration area.

OMA used the average concentration over the whole exploration area rather than that for the mineable area only (Ocean Mining Associates, 1982). OMCO defines the value as the "average abundance over *acceptable* area, kg/m^2" without defining "acceptable" (Ocean Minerals Company, 1982).

It is reasonable to conclude that the size of an exploration area will be at least twice the eventual area to be mined; it is also possible that it will be several times larger. It is also important to keep in mind that each area will differ to some extent and the ratio of exploration area to mining area will differ accordingly.

EXPLORATION AREAS LICENSED UNDER US DOMESTIC LAW

As part of the procedures followed by NOAA during the drafting of regulations for exploration licences, public comment was solicited on issues related to the regulations. Several proposals on the size of exploration areas arose during this process. An early draft of the regulations called for an area size limit of 80,000 km^2. This was deleted from the final rules after it met with objections (NOAA, 1981a). In a discussion paper prepared by NOAA (1980), it was suggested that "on the assumption of 50 percent unmineable ground and compatible with the proposed Law of the Sea Treaty, allowing the proposal of two mine-sites, an exploration area might reasonably be expected to total 120,000–160,000 km^2". It is unclear from the text whether an exploration area in this size range was intended to eventually allow one or two mining operations.* However, in the original applications made to NOAA in 1982, each consortium requested at least two exploration areas of sizes in keeping with those agreed upon in the Convention of the Law of the Sea. As shown in Table 10, some were exactly 150,000 km^2. Two years later, in the final applications, after settlement of conflicts arising from overlapping claims, the sizes of licence areas granted were as follows: KCON, 65,000 km^2; OMA, 156,000 km^2; OMI, 136,000 km^2; OMCO, 165,533 km^2. In the final regulations, the applicant is requested to justify through logical arguments and with technical support, that the size of the exploration area is reasonable. Thus, no fixed exploration area size is required.

In late 1984, the four mining consortia who were issued licences under US national legislation by NOAA requested that the geographical coordinates delineating the areas be made public. (See Appendix C, Figures and Tables.) Figure 6 shows the US licensed areas plotted in the Clarion-Clipperton Zone. Missing from Fig. 6 are four other areas for which overlap

*Under the Convention on the Law of the Sea, an applicant must submit a plan of work for an area large enough to include two mining operations. When approval is given, the area is split into two halves of equal estimated commercial value.

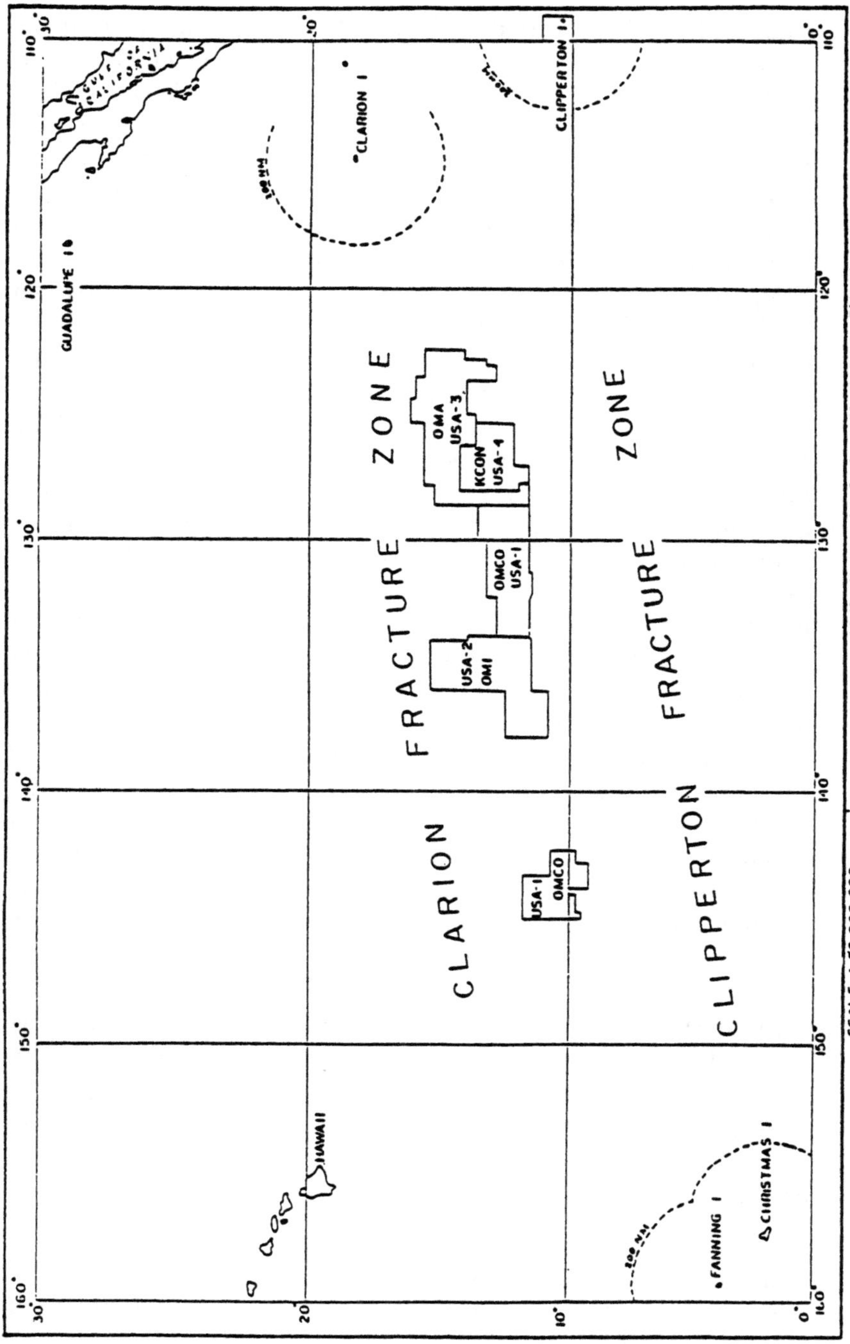

Fig. 6. Exploration Areas Licensed under U.S. national legislation.

conflicts have been settled. OMI has filed a second application for an exploration area with the Federal Republic of Germany; KCON has filed a second licence application with the United Kingdom; Japan and France have submitted applications to the Preparatory Commission for the International Sea-Bed Authority and for the International Tribunal for the Law of the Sea of the United Nations. In addition to the above AMR has reportedly applied for a site under FRG legislation outside the Clarion-Clipperton Zone and the Soviet Union has also submitted an application to the Preparatory Commission for a site in the Clarion-Clipperton Zone. Other anticipated sites are outside the Clarion-Clipperton Zone; the Indian government has submitted an application to the Preparatory Commission for a site in the Indian Ocean. The Japanese, French and Soviet applications under the Law of the Sea Convention could involve areas as large as 300,000 km^2 each. The United Kingdom KCON site is approximately 118,000 km^2. OMI originally applied for two licences from NOAA, the total area of which was 559,000 km^2. It is not known at this time how large an area was applied for in the Federal Republic of Germany since coordinates are proprietary information and are not published in the FRG. Summing the total amount of areas already licensed under US law with the maximum amount of area allowed to applicants under the Law of the Sea Convention, a maximum of approximately 2 million km^2 could conceivably be awarded as exploration areas. However, the applicants for registration with the United Nations Preparatory Commission have also engaged in negotiations to settle conflicting claims due to overlap (United Nations, 1986). The anticipated result of these negotiations will undoubtedly result in a reduction of area.

From Fig. 6 and from the fact that there are still conflicts to be resolved, two observations can be made with regard to the availability of the resource for first-generation mining: there is an intense concentration of interest in a relatively small area within the Clarion-Clipperton Zone and paramarginal resources are limited almost exclusively to that area.

---

*This estimate is approximately the same as that obtained by other researchers whose considerations were based on other factors such as the requirements of production limitation from the Area (Lèvy and Odunton, 1984).

Based on the total number of exploration areas that have been awarded or can be expected to be awarded, it appears that there will be about 9 to 12 first-generation mining projects, with perhaps three outside the Clarion-Clipperton Zone.* The total amount of area awarded for exploration could be between 1 and 1.5 million square kilometres.

Table 10. Exploration area sizes in 1982 applications to NOAA

Applicant	*Exploration Area Size* (km^2)
KCON	150,000
KCON	150,000
OMCO	119,351
OMCO	118,209
OMI	136,000
OMI	130,000
OMI	149,000
OMI	144,000
OMA	150,000
OMA	150,000

Chapter 5

MINE SITE CONFIGURATION

In 1974 the United Nations prepared a technical document showing different possible configurations for ocean mine-sites. While the options have since narrowed, the concepts relating to the configurations are still of interest. This chapter reviews the different possible configurations for ocean mine sites, examines the current discussion of the issues they raise and points out some of the implications inherent in the different approaches.

HYPOTHETICAL APPROACHES

Based on the terminology used, the earliest suggestions about mine-site configuration apparently stemmed from prior experience with offshore oil leasing. The 1975 US Senate Bill S.713, defined areas as "blocks" of 40,000 km^2, the width of which could not be less than one-sixth the length.

The same idea is evident in a United Nations document (A/CONF.62/25) which presented the idea of a hypothetical master grid of the International Area made up of blocks 50 km×50 km (2,500 km^2). It also suggested the possibilities of a grid with a global character with all blocks similarly defined by geographic coordinates, and a grid of a regional character, where the size of blocks would vary according to geological characteristics of the seafloor. Another alternative was to allow the nodule miners to define the block size and shape themselves. The hypothetical result of this alternative is portrayed in Fig. 7.

The United Nations recognized the problem of determining an appropriate block size in that the quality and quantity of the resource can vary from site to site. In addition, the authors cited the problem of variations in the efficiencies of the recovery systems. Given these factors the United Nations document lists three alternative approaches to mine-site size (and by implication, shape): (a) variable sizes as requested by the miner; (b) a

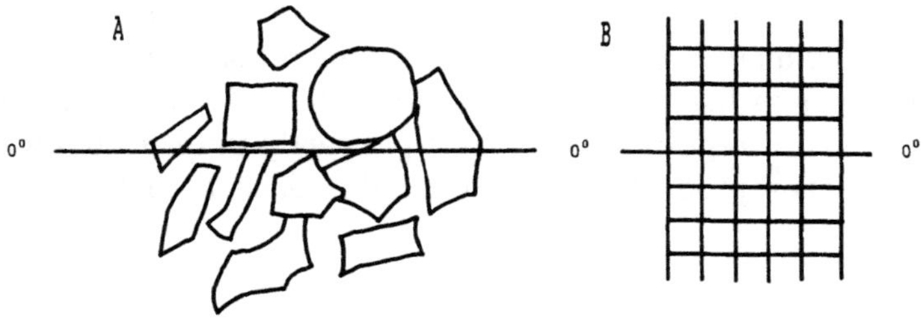

Fig. 7. Two hypothetical cases of subdivision of the International Area. From: United Nations, A/Conf. 62/25 (Fig. 9).

standard size for all blocks regardless of locality and proposed production capacity; and (c) variable sizes predetermined by a master grid.

Another model of a hypothetical mine-site that combines features mentioned by the United Nations paper is presented in a discussion of environmental monitoring alternatives as shown in Fig. 8 (Andrews *et al.*, 1983). These configurations combine the following elements: (a) size and shape are variable and determined by the miner; and (b) the sites are delineated by latitude and longitude, resulting in rectangular shapes, not irregular geometric shapes. Some unmineable area, presumably as little as possible, would be included in the final site determination.

EVOLVING REGULATORY APPROACHES

Under Resolution II of the United Nations Convention on the Law of the Sea, the upper limit on exploration area size was set at 150,000 km^2. A further requirement compels the operator to relinquish over a period of time at least half of the area that was agreed upon for exploration. Under the Convention itself, no fixed area size for either exploration or exploitation was adopted. Thus, under the Convention, it is possible that a system of individually calculated mine-site sizes could be adopted.

As for shape, under the Law of the Sea Convention, it is stated that the area "need not be a single continuous area" (Annex III, Article 8). This will allow for configurations such as those depicted in Fig. 8 or for several areas separated by even greater distances as shown in Fig. 9.

In developing mining regulations under domestic law, the

United States has moved from an initial concept of fixed area size to an individually calculated size for exploration area. As for the mine-site area itself, or the "logical mining unit" as it is referred to, the permit regulations define it primarily in terms of an area that satisfies production requirements. Each area will be ex-

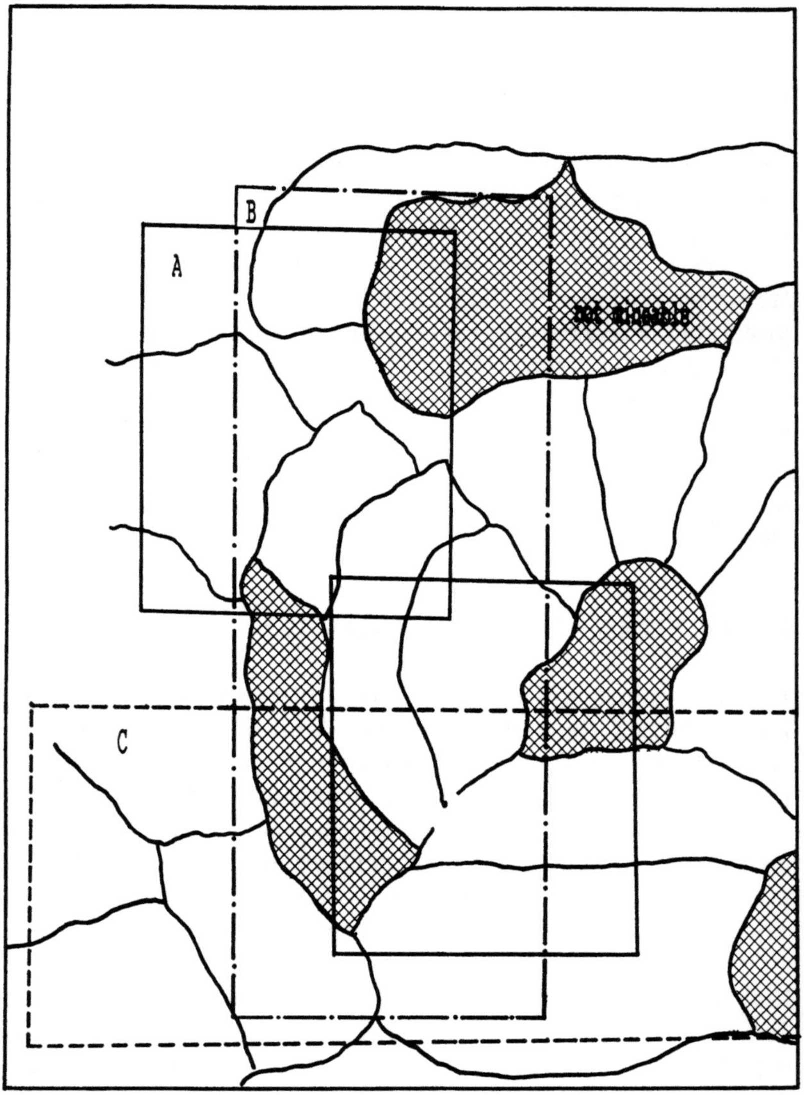

Fig. 8. Three possible mine site configurations. From Andrews *et al.*, 1983 (Fig. 17).

amined individually and need not be a single continuous segment as long as each segment can be mined efficiently (NOAA, 1986).

What are the implications for management of the resource under these two regulatory approaches?

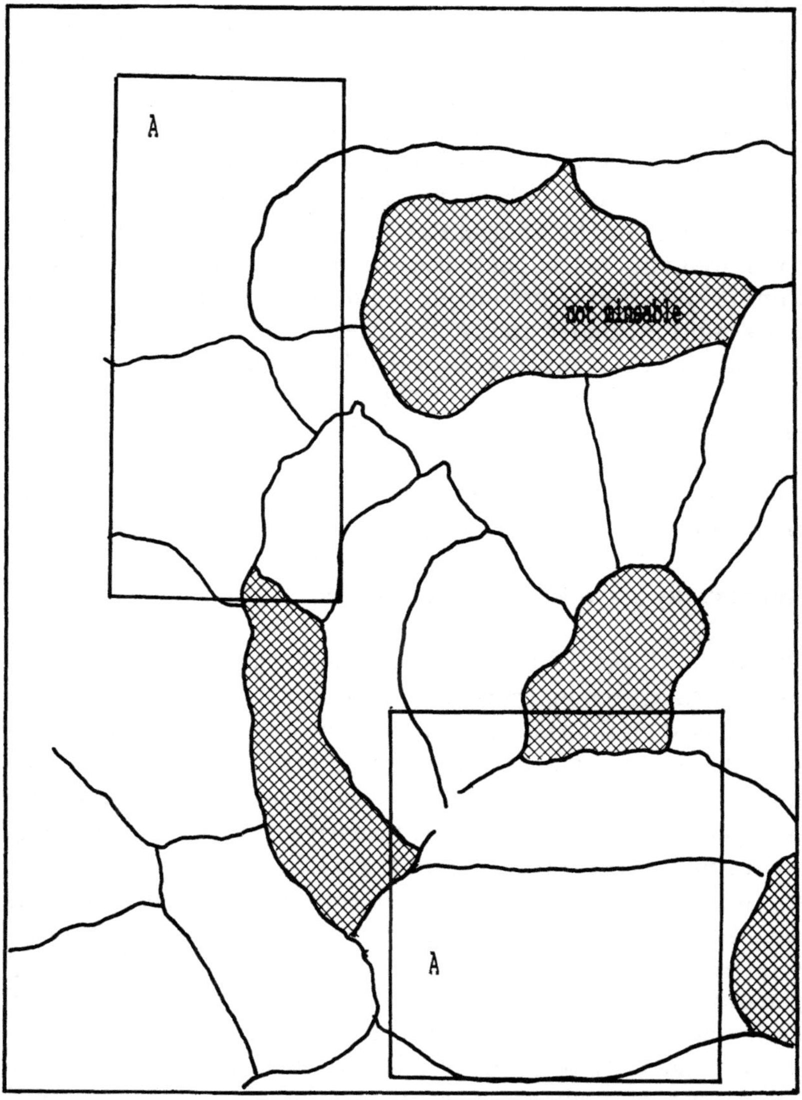

Fig. 9. Possible non-continuous mine-site configurations. After Andrews *et al.*, 1983 (Fig. 17).

Fixed Area Sizes (LOS)

A large fixed area size when looked at from one perspective has the immediate advantage of minimal administrative requirements. Less detailed information on the site itself and on system factors would be required from the operator since no assessment of that information would be necessary for an area size award. This approach would also almost certainly decrease the number of near-term viable mine-sites and logically the number of operators. This situation favours the first-generation operators and minimizes the need for increasing the efficiency of the mining system. Since size estimates made in 1982 were based on 1982 projected efficiency values, various situations could be anticipated. For example, an operator with exclusive rights to a large fixed area may not be motivated to increase the efficiency of his mining system since production requirements could be met on the basis of his estimated 1982 efficiencies regardless of what year he actually begins operation. On the other hand, an operator could increase his efficiency by, say, the year 2000 and could increase significantly his production from the area awarded. This, of course, would in part be determined by whether the duration of the operation is fixed or subject to renewal conditions, and, under the Law of the Sea Convention, by the production limitation.

Individually Sized Mine-site Areas (US)

While, in theory, this approach should lead to making more mine-sites available, there is no guarantee that this will be the case. Based on information given by the operator, and assuming that there is no independent verification of the information, the mine-site could be whatever size the operator deems necessary for his operation. The same lack of incentive to improve efficiencies would hold as under the fixed area size scheme. NOAA states that its position is one which "intends to encourage the wise use of the deep seabed manganese nodule resource so as to insure that its role as a resource for the future is not precluded by first generation mining" (NOAA, 1982). NOAA has in fact indicated a preference in its Programmatic Environmental Impact Statement to "identify the logic of selective mining plan" in order to avoid a situation where "the applicant could mine rich zones until leaner ones are all that remain and then abandon the site" (NOAA, 1981b). However, identifying the logic of a selective mining plan does not preclude high grading the deposit if the logic is convincing.

Part of the willingness to allow such latitude to the miner from both the international community and NOAA is clearly linked to the pioneer nature of the seabed mining industry. Given the uncertainties and the unknowns related to a technological configuration that has yet to be proven, both approaches reflect a willingness to encourage the development of a promising source of raw materials.

Chapter 6

CONCLUSIONS

A manganese nodule mine-site is an area that meets certain minimum geological, economic and technical conditions, the most important of which, with their present and likely values, are:

Cut-off grade	1.8 percent Cu plus Ni
Cut-off abundance	5 kg/m^2
Topography	acceptable
Duration of recovery	20 years
Annual recovery	3×10^6 dry metric tons
Overall efficiency of mining system	approx. 20 percent

The size of a mine-site will depend on the combination of these factors. The mine-site will consist of mineable areas the sizes of which will vary from approximately 10 km^2 to 50 km^2. Each field will have to meet the requirements of a specified mining system. For the first generation of mining systems these requirements could include conditions such as: (i) no slopes within mineable areas that are steeper than a ten percent gradient and, (ii) no obstacles within mineable areas that can either not be avoided by the dredge head or detected by it.

The mineable areas will be several kilometers apart and look like patches of colour on a turkish carpet. From one claim site to another the spacing between mineable areas will change. These changes, however, will not alter the profitability of a given operation because the time lost in transit from one mineable area to another within a claim site will generally be insignificant when compared with the time required to mine each mineable area. The range of mine-site sizes extends from about 15,000 km^2 to 125,000 km^2. Exploration areas will be at least twice the anticipated mine-site size.

Based on the estimated availability of the resources, the number of first-generation mine-sites maybe between 10 and 30 in the Clarion-Clipperton Zone: however, it is more likely that there will be about 9–12 first generation mining operations in view of the small number of qualified operators and the relatively limited area of the ocean containing paramarginal resources.

Appendix A

Definition of Terms

It has often been noted that the literature on seabed mining is filled with a confusing and mixed use of terms to describe many of its elements. The following list of selected terms is presented along with their definitions in order to provide a clear and consistent basis for the discussion that follows.

Abundance (also, nodule density, nodule concentration) is the weight of nodules per unit area of seafloor, usually expressed as kg/m^2 either wet or dry. Dry abundance is about 70 percent of wet abundance as nodules are extremely porous and considered to contain about 30 percent water.

Cut-off abundance (also, adequate abundance, minimum average abundance) is the minimum average abundance in the areas that can be mined to satisfy production requirements.

Cumulative recovery is the total quantity of nodules to be recovered from an area. It is determined by economic considerations and is usually expressed as (dry) metric tons.

First-generation mining is mining using the first-generation of technology. For additional information on first-generation mining and processing technologies, see Volumes 2 and 3 of this Series.

Grade is the content of an element of commercial interest and is usually expressed as a percentage of the dry weight. In seabed mining literature, grade is used to denote individual values for copper, nickle, cobalt and manganese, or, more frequently, the sum of copper plus nickle.

Cut-off grade (also, adequate grade) is defined: (1) as the minimum grade that could be profitably recovered, and (2) as the minimum related to the required average grade.

Average grade (also, required average grade, minimum average grade) is the mean value as calculated from the class intervals of an assumed grade distribution that meet or exceed the criteria for cut-off grade.

In practice, minimum average grade and cut-off grade are both used to delineate elementary blocks which cannot be divided during mining. The useful economic parameter is average grade;

cut-off grade is the lowest grade included in the average. Cut-off grade is used, therefore, as a simple lower limit to enable computation and contouring.

Mineable area (also, mineable proportion) is the part of the area selected in which mining would actually be attempted.

Net mining efficiency (also, overall mining recovery efficiency, overall efficiency, mining efficiency) is the percentage of nodules in the total mine-site area that may be actually recovered.

Appendix B

Acronyms of Groups Involved in Mine-site Development

AFERNOD	Association Francaise pour l'Etude et la Recherche des Nodules
AMR	Arbeitsgemeinschaft Gweinnbare Rohstoffe
CCOP/SOPAC	Co-ordination of Joint Committee for Prospecting for Mineral Resources in South Pacific Offshore Areas
CNEXO	Centre National pour l'Exploitation des Océans (now IFREMER)
DOMA	Deep Ocean Minerals Association
DORD	Deep Ocean Development Co.
DSV	Deepsea Ventures, Inc.
IFREMER	Institut Francaise de Recherche pour l'Exploitation de la Mer
KCON	Kennecott Consortium
OMA	Ocean Mining Associates
OMCO	Ocean Minerals Company
OMI	Ocean Management, Inc.

For additional information on structure and composition of some of the above groups, see Volume 2 of this Series.

Appendix C

Maps and Turning Points of US Licensed Exploration Areas

Kennecott Consortium (KCON)

Turning Point	Latitude	Longitude
1	14°20' N	128°00' W
2	14°20' N	126°15' W
3	13°45' N	126°15' W
4	13°45' N	125°20' W
5	12°15' N	125°20' W
6	12°15' N	127°00' W
7	11°40' N	127°00' W
8	11°40' N	127°43' W
9	12°00' N	127°43' W
10	12°00' N	128°00' W
1	14°20' N	128°00' W

KCON 65,000km^2

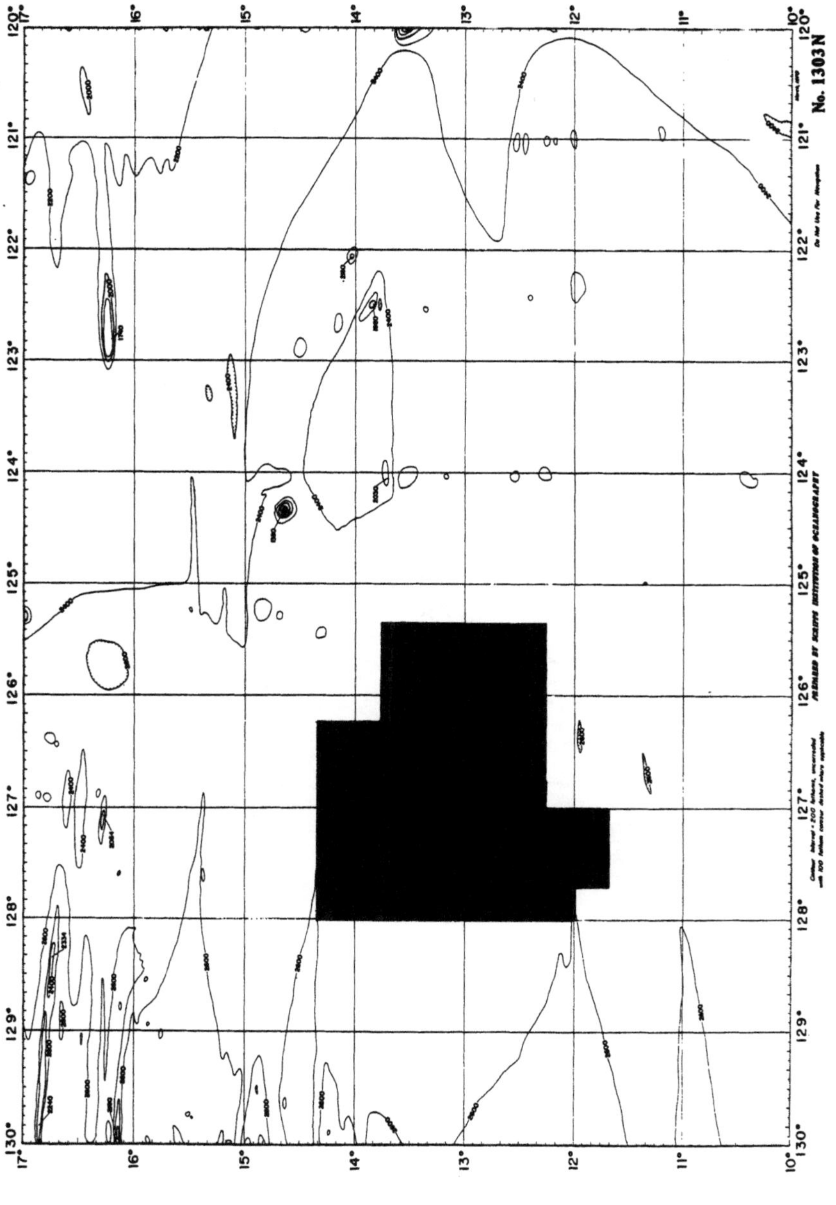

Ocean Mining, Inc. (OMI)

Turning Point	Latitude	Longitude
1	15°25' N	134°00' W
2	14°00' N	134°00' W
3	14°00' N	133°50' W
4	11°30' N	133°50' W
5	11°30' N	136°00' W
6	10°50' N	136°00' W
7	10°50' N	137°50' W
8	12°30' N	137°50' W
9	12°30' N	136°00' W
10	15°25' N	136°00' W
1	15°25' N	134°00' W

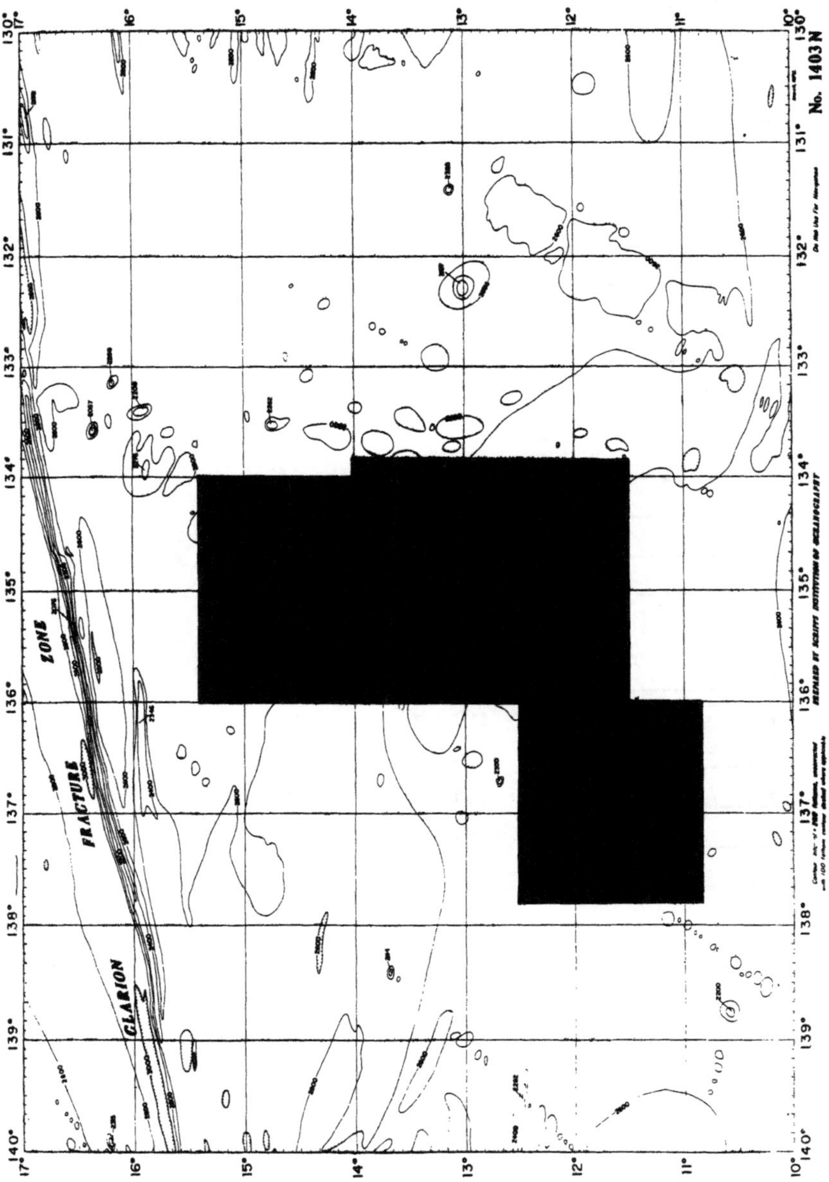
CLARION
FRACTURE
ZONE
No. 1403 N

Ocean Minerals Company (OMCO)

Turning Point	Latitude	Longitude
Area 1:		
1	13°40' N	128°35' W
2	11°40' N	128°35' W
3	11°40' N	131°15' W
4	11°30' N	131°15' W
5	11°30' N	132°00' W
6	11°40' N	132°20' W
7	11°40' N	133°50' W
8	12°50' N	133°50' W
9	12°50' N	132°15' W
10	13°20' N	132°15' W
11	13°20' N	130°00' W
12	13°40' N	130°00' W
1	13°40' N	128°35' W
Area 2		
1	11°50' N	145°00' W
2	11°50' N	143°15' W
3	10°45' N	143°15' W
4	10°45' N	142°15' W
5	9°45' N	142°15' W
6	9°45' N	142°45' W
7	9°15' N	142°45' W
8	9°15' N	143°45' W
9	10°00' N	143°45' W
10	10°00' N	144°00' W
11	9°45' N	144°00' W
12	9°45' N	144°45' W
13	9°30' N	144°45' W
14	9°30' N	145°00' W
1	11°50' N	145°00' W

Area 1 103,190km

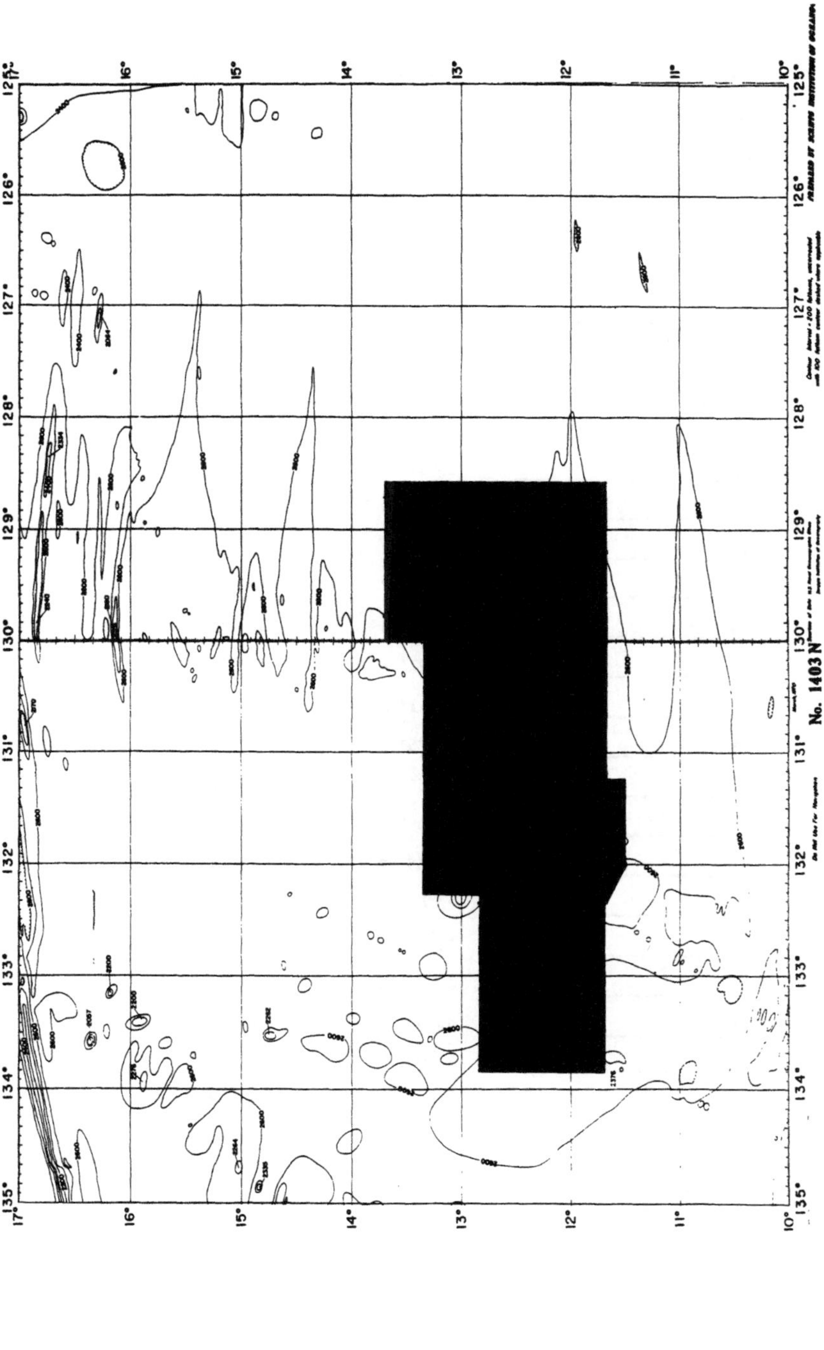

Ocean Mining Associates (OMA)

Turning Point	Latitude	Longitude
1	15°20' N	128°35' W
2	15°20' N	127°50' W
3	15°15' N	127°50' W
4	15°15' N	127°46' W
5	15°44' N	127°46' W
6	15°44' N	125°20' W
7	16°14' N	125°20' W
8	16°14' N	124°20' W
9	16°04' N	124°20' W
10	16°04' N	123°25' W
11	15°44' N	123°25' W
12	15°44' N	122°20' W
13	14°10' N	122°20' W
14	14°10' N	122°45' W
15	13°21' N	122°45' W
16	13°21' N	123°00' W
17	12°56' N	123°00' W
18	12°56' N	123°35' W
19	14°05' N	123°35' W
20	14°05' N	125°00' W
21	13°45' N	125°00' W
22	13°45' N	126°15' W
23	14°20' N	126°15' W
24	14°20' N	128°00' W
25	12°00' N	128°00' W
26	12°00' N	127°43' W
27	11°40' N	127°43' W
28	11°40' N	128°35' W
1	15°20' N	128°35' W

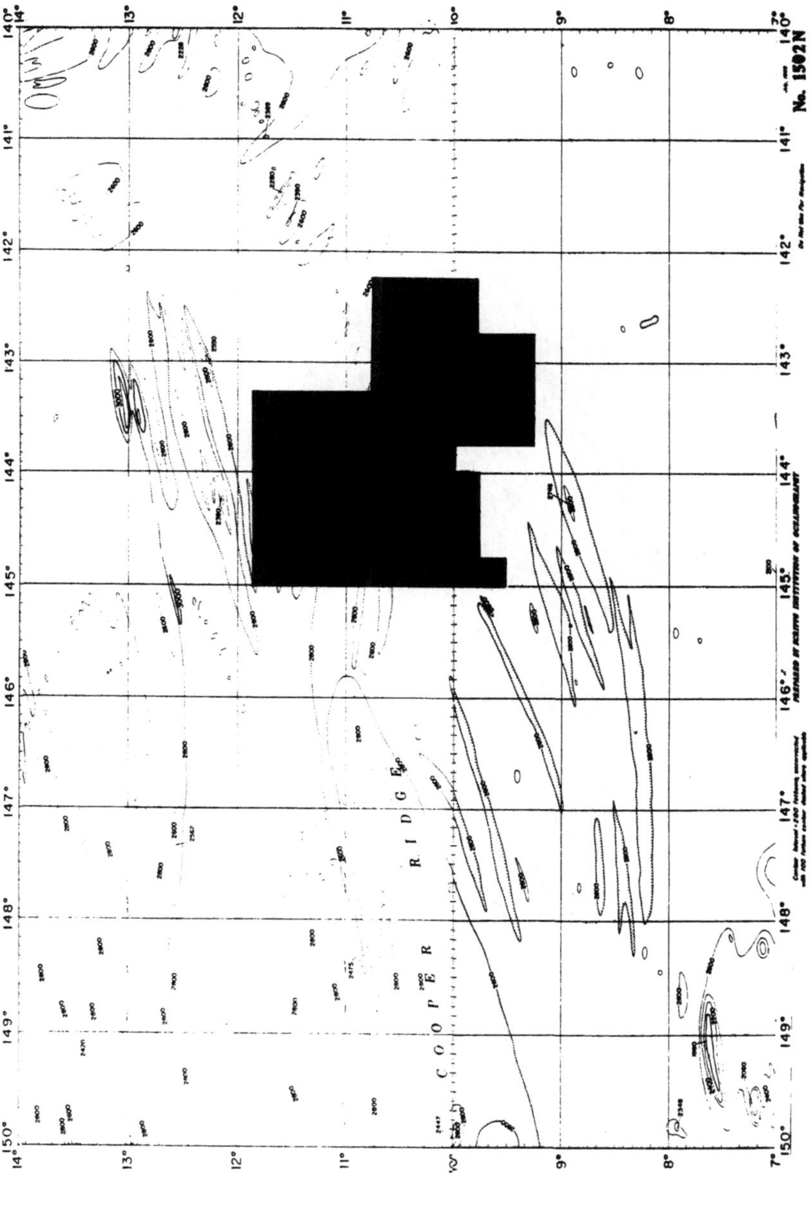

OMCO Area 2 62,343km^2

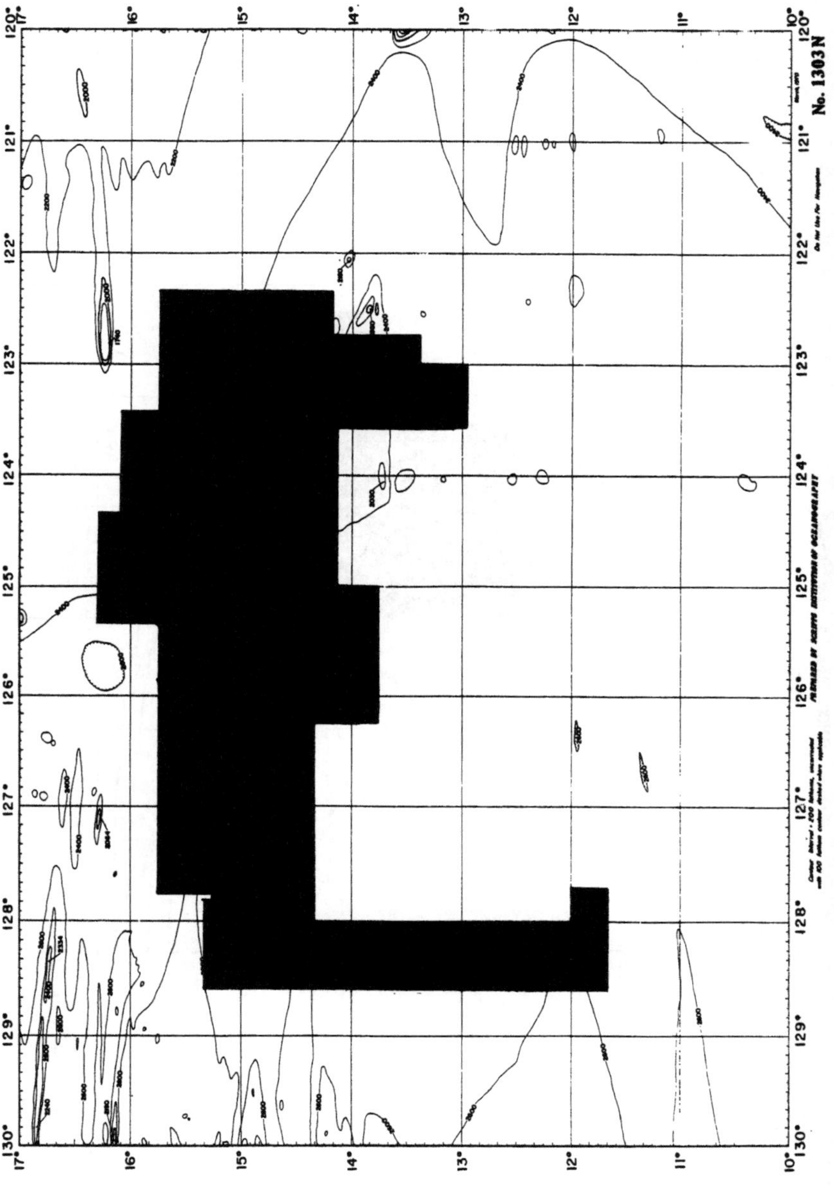

OMA 156,000km^2

References

l'Académie des Sciences, 1984. *Les Nodules polymétalliques. Faut-il exploiter les mines océaniques?* Gauthier-Villars (Paris), 180 pp.

Amann, H. M., 1975. Definition of an ocean mining site. Offshore Technology Conference, *Proceedings,* Vol. I, OTC 2238, pp. 891–908.

Andrews, B. V., J. E. Flipse and F. C. Brown, 1983. *The Economic Viability of a Four-Metal Deep Ocean Mining Venture.* Texas A & M University, TAMU-SG-84-201, October, 201pp.

Archer, A. A., 1976. Prospects for the exploitation of manganese nodules: the main technical, economic and legal problems. Papers presented at the I.D.O.E. Workshop, Suva, Fiji, 1975, eds. Glasby, G. P. and H. R. Katz, *Technical Bulletin No. 2*, CCOP/SOPAC, pp. 21–38.

Archer, A. A., 1979. Resources and potential reserves of nickle and copper in manganese nodules. United Nations Ocean Economics and Technology Office, *Manganese Nodules: Dimensions and Perspectives*, pp. 71–81.

Archer, A. A., 1981. Manganese nodules as a source of nickle, copper, cobalt and manganese. *Trans. Instn. Min. Metall. (Sect. A: Min. industry),* **90**, January, pp. A1–A6.

Bastien-Thiry, H., J-P, Lenoble and P. Rogel, 1977. French exploration seeks to define mineable nodule tonnages on Pacific floor. *Eng. and Min. J1.*, July.

Bastien-Thiry, H. 1979. Sampling and surveying techniques. United Nations Ocean Economics and Technology Office, *Manganese Nodules: Dimensions and Perspectives*, pp. 7–19.

Charles River Associates, Inc., 1982. Ocean mine-sites. In *Analysis of Major Policy Issues Raised by the Commercial Development of Ocean Manganese Nodules*, July.

Deepsea Ventures, Inc., 1974. *Notice of Discovery and Claim of Exclusive Mining Rights and Request for Diplomatic Protection of Investment,* filed November 1974 with the U.S. Secretary of State, U.S. Department of State, Washington, D.C.

ESCAP/CCOP/SOPAC, 1981. Prospecting for manganese nodules by France in the South Pacific. NR/CCOP/SOPAC (10) CR.7-I-51, submitted by French delegation at Tenth Session, 6–14 October, Port Vila, Vanuatu.

Exon, N. F., 1983. Manganese nodule deposits in the central Pacific Ocean and their variation with latitude. *Marine Mining*, **4**, No. 1, pp. 79–107

Fellerer, R., 1975. Methods and problems of the exploration of manganese nodules. *Oceanology International,* **75**, pp. 121–126.

Fisk, M. B., J. Z. Frazer, J. S. Elliott and L. L. Wilson, 1979. Availability of copper, nickel, cobalt and manganese from ocean ferromanganese nodules (II). Scripps Institution of Oceanography, Ref. 79–17, August 24, 68 pp.

Fisk, M. B., 1982. Seabed minerals. Presented at Law of the Sea Seminar, Nov. 15–16, Pacific Islands Assoc., New York (unpublished).

Flipse, J. E., Dubs, M. A. and R. J. Greenwald, 1973. Pre-production manganese nodules activities and requirements. Mineral Resources of the Deep Sea-bed, Hearings before the Sub-committee on Minerals, Materials and Fuels of the U.S. Senate Committee on Interior and Insular Affairs, 15 March, pp. 602–700.

Frazer, J. Z., 1977. Manganese nodule reserves: an updated estimate. *Marine Mining,* **1**, No. 1/2, pp. 103–123.

Frazer, J. Z. and M. B. Fisk, 1978. *Sediment Data Bank User's Handbook.* Scripps Institution of Oceanography Ref. 78–10, May, 74 pp.

Frazer, J. Z., M. B. Fisk, J. Elliott, M. White and L. Wilson, 1978. Availability of copper, nickle, cobalt and manganese from ocean ferromanganese nodules. Scripps Institution of Oceanography, Ref. 78-25, September 10, 139 pp.

Frazer, J. Z., 1979. The reliability of available data on element concentrations in seafloor manganese nodules. United Nations Ocean Economics and Technology Office, *Manganese Nodules: Dimensions and Perspectives,* pp. 21–36.

Frazer, J. Z., 1980. Resources in seafloor manganese nodules. In *Deepsea Mining*, ed. J. T. Kildow, (MIT Press), Cambridge, Mass. London, pp. 41–83.

Frazer, J. Z. and M. B. Fisk, 1980. Availability of copper, nickle, cobalt and manganese from ocean ferromanganese nodules (III). Scripps Institution of Oceanography, Ref. 80-16, August 31, 117 pp.

Frazer, J. Z. and L. L. Wilson, 1980. Manganese nodule resources in the Indian Ocean. *Marine Mining*, **2**, No. 3, pp. 257–292.

Glasby, G. P., 1983. The three-million-tons per year manganese nodule mine-site: an optimistic assumption? *Marine Mining,* **4**, No. 1, pp. 73–77.

Glasby, G. P., 1984. Manganese nodule research in the Federal Republic of Germany: a review. *Marine Mining,* **4**, No. 4, pp. 355–402.

Geological Survey of Japan, 1983. Manganese Nodules Distribution in the Central Pacific Ocean. Marine Geology Map Series 21.

Halbach, P. and R. Fellerer, 1980. The metallic minerals of the Pacific seafloor. *Geojournal,* **4**, pp. 407–422.

Halbach, P., 1984. Deep-sea metallic deposits. *Ocean Management,* **9**, pp. 35–60.

Halkyard, J. E., 1979. Ocean mining technology and environmental studies. *Mining Congress J1.*, **65**(8).

Hillman, C. T. and R. D. Weldin, 1976. A preliminary evaluation of a deep ocean polymetallic nodule deposit. Unpublished U.S. Bureau of Mines report, available for consultation at Western Field Operations Centre, Spokane, WA, 38 pp.

Hillman, C. T., 1983. Manganese nodule resources of three areas in the northeast Pacific Ocean: with proposed mining-beneficiation systems and costs. A minerals availability system appraisal. U.S. Dept. of the Interior, Bureau of Mines Information Circular, IC 8933, 60 pp.

Holser, A. F., 1976. Manganese nodule resources and mine-site availability: professional staff study, Ocean Mining Administration, U.S. Dept. of the Interior, Washington, D.C., 12 pp.

Horn, D. R., B. M. Horn and M. N. Delach, 1973. *Ocean Manganese Nodules Metal Values and Mining Sites*. Lamont-Doherty Geological Observatory of Columbia University, Tech. Rep. No. 4, 57 pp.

Kaufman, R., 1974. The selection and sizing of tracts comprising a manganese nodule ore body. Offshore Technology Conference, Preprints, Vol. II, OTC 2059, pp. 283–293.

Lenoble, J-P., 1980. Polymetallic nodules resources and reserves in the north Pacific from the data collected by AFERNOD. *Oceanology International*, **80**, pp. 11–18.

McKelvey, V. E., N. A. Wright and R. W. Rowland, 1979. Manganese nodule resources in the northeastern equatorial Pacific. In: *Marine Geology and Oceanography of the Pacific Manganese Nodule Province*, eds. Bischoft, J. L. and D. Z. Piper, (Plenum), pp. 747–757.

McKelvey, V. E., N. A. Wright and R. W. Bowen, 1982. Analysis of the world distribution of metal-rich subsea manganese nodules. U.S. Dept. of the Interior, Geological Survey Circular 886, 55 pp.

Menard, H. W. and J. Z. Frazer, 1978. Manganese nodules on the sea floor; inverse correlation between grade and abundance. *Science*, **199**, 3 March, pp. 969–971.

National Oceanic and Atmospheric Administration, 1980. Discussion Paper on Implementation of Public Law No. 96-283, The Deep Seabed Hard Minerals Resources Act, Nov., 66 pp.

National Oceanic and Atmospheric Administration, 1981a. Deep Seabed Mining Regulations for Exploration Licenses: Final Rules, Part IV, Federal Register, Vol. 46, No. 178, Sept. 15, pp. 45890–45920.

National Oceanic and Atmospheric Administration, Office of Ocean Minerals and Energy, 1981b. Deep Seabed Mining Programmatic Environmental Impact Statement, Vol. I, 295 pp.

National Oceanic and Atmospheric Administration, 1982. Deep Seabed Mining: Regulations for Commercial Recovery; Advanced Notice of Proposed Rulemaking, Federal Register, Vol. 47, No. 249, Dec. 28, pp. 57904–57906.

National Oceanic and Atmospheric Administration, 1986. Deep Seabed Mining; Commercial Recovery and Exploration Regulations; Proposed Rule. Federal Register, Vol. 51, No. 1143, Friday, July 25, pp. 26794–26824.

Ocean Management, Inc., 1982. *Application for and Notice of Claim to Exclusive Exploration Rights for Manganese Nodule Deposits in the Northeast Equatorial Pacific Ocean,* filed with the National Oceanic and Atmospheric Administration, U.S. Department of Commerce, 19 February.

Ocean Minerals Company, 1982. *Ocean Minerals Company Deep Seabed Mining Exploration License Applications,* filed with the National Oceanic and Atmospheric Administration, U.S. Department of Commerce, 25 January.

Ocean Mining Associates, 1982. *Application by Ocean Mining Associates for an Exploration License,* filed with the National Oceanic and Atmospheric Administration, U.S. Department of Commerce, 18 February.

Pasho, D. W. and J. A. McIntosh, 1976. Recoverable nickle and copper from manganese nodules in the northeastern equatorial Pacific — preliminary results. *Canadian Inst. of Min. and Metal. Bull.,* **69**, pp. 15–16.

Pasho, D. W. 1977. Review of the development of deep seabed manganese nodules. *The Northern Miner,* April 14, pp. B6, B9, B13.

Pasho, D. W., 1979. Determining deep sea-bed mine-site area requirements — a discussion. United Nations Ocean Economics and Technology Office, *Manganese Nodules: Dimensions and Perspectives,* pp. 83–112.

Siapno, W. D., 1975. Exploration technology and ocean mining parameters. *American Mining Congress Convention*, San Francisco, 28 September–1 October, reprint, 24pp.

U.S. Congress, 1974. House Committee on Merchant Marine and Fisheries. Deep seabed hard minerals. Hearings before the Subcommittee on Oceanography on H.R. 9 and H.R. 7732. 93rd Cong. Mar 1, 28–29, April 3, 1973, H.R. 12233, Feb. 26, 27, 28, U.S. Gov't Printing Office, Washington, D.C., 1974, 513 pp.

U.S. Congress, 1975. Senate. Committee on Interior and Insular Affairs. Current Developments in Deep Seabed Mining. Hearing before the Subcommittee on Minerals, Materials and Fuels. 94th Congress. 1st sess., Nov. 7, Washington, D.C., U.S. Gov't Print. Off., Part I, 249pp.

United Nations, Third Conference on the Law of the Sea, 1974. Economic Implications of Seabed Mineral Development in the International Area. Report of the Secretary-General, A/CONF.62/25, May 22, 80pp.

United Nations, 1979. *Manganese Nodules: Dimensions and Perspectives.* D. Reidel (Dordrect, Holland) 194pp.

United Nations, 1982. *Assessment of Manganese Nodule Resources: The Data and the Methodologies.* Graham and Trotman (London), 79 pp. Seabed Minerals, Vol. 1.

United Nations, 1983. *The Law of the Sea.* Official Text of the United Nations Convention on the Law of the Sea with Annexes and Index, United Nations, New York.

United Nations, 1986. Statement by the Chairman of the Preparatory Commission. Preparatory Commission for the International Sea-Bed Authority and for the International Tribunal for the Law of the Sea, Geneva, 12 August–4 September, 1985, LOS/PCN/L.27/Rev.1.

List of Figures

List of Tables

Index

GPSR Compliance

The European Union's (EU) General Product Safety Regulation (GPSR) is a set of rules that requires consumer products to be safe and our obligations to ensure this.

If you have any concerns about our products, you can contact us on ProductSafety@springernature.com

In case Publisher is established outside the EU, the EU authorized representative is:

Springer Nature Customer Service Center GmbH
Europaplatz 3
69115 Heidelberg, Germany

Batch number: 10607194

Printed by Printforce, the Netherlands